Palgrave Macmillan Memory Studies

The nascent field of Memory Studies emerges from contemporary trends that include a shift from concern with historical knowledge of events to that of memory, from 'what we know' to 'how we remember it'; changes in generational memory; the rapid advance of technologies of memory; panics over declining powers of memory, which mirror our fascination with the possibilities of memory enhancement; and the development of trauma narratives in reshaping the past. These factors have contributed to an intensification of public discourses on our past over the last thirty years. Technological, political, interpersonal, social and cultural shifts affect what, how and why people and societies remember and forget. This groundbreaking series tackles questions such as: What is 'memory' under these conditions? What are its prospects, and also the prospects for its interdisciplinary and systematic study? What are the conceptual, theoretical and methodological tools for its investigation and illumination?

Simon Lewis • Jeffrey Olick
Joanna Wawrzyniak • Malgorzata Pakier
Editors

Regions of Memory

Transnational Formations

palgrave
macmillan

Editors
Simon Lewis
University of Bremen
Bremen, Germany

Jeffrey Olick
Department of Sociology and History
University of Virginia
Charlottesville, VA, USA

Joanna Wawrzyniak
Faculty of Sociology, Center for Research on Social Memory
University of Warsaw
Warsaw, Poland

Malgorzata Pakier
European Network Remembrance and Solidarity
Warsaw, Poland

ISSN 2634-6257 ISSN 2634-6265 (electronic)
Palgrave Macmillan Memory Studies
ISBN 978-3-030-93704-1 ISBN 978-3-030-93705-8 (eBook)
https://doi.org/10.1007/978-3-030-93705-8

This Palgrave Macmillan imprint is published by the registered company Springer Nature Switzerland AG.
The registered company address is: Gewerbestrasse 11, 6330 Cham, Switzerland

Acknowledgments

The editors are grateful to the European Network Remembrance and Solidarity for generously supporting two conferences at which the main idea of this book originated: "Regions of Memory: A Comparative Perspective on Eastern Europe" (2012) and "Regions of Memory II: Memory Regions as Discourse and Imagination" (2016). We would also like to thank all of the participants of both conferences, as the many contributions and comments helped us to reconsider and refine our approach. We extend our appreciation to the two anonymous reviewers for their helpful feedback, Elsbeth van der Wilt for her invaluable work in copyediting, as well as to the Palgrave editorial team for their professional assistance. The publication of this book was supported by a grant from the University of Warsaw in the framework of the program "Excellence Initiative—Research University", as well as by the ECHOES project, which received funding from the European Union's Horizon 2020 research and innovation program under grant agreement No. 770248.

Contents

Notes on Contributors

Tony Joel is Associate Professor in History at Deakin University. A former German Academic Exchange Service (DAAD) scholarship holder and a member of Deakin's Contemporary Histories Research Group, Tony's main research interests include war memory and commemoration, and sports history. Tony's publications include *The Dresden Firebombing: Memory and the Politics of Commemorating Destruction* (Bloomsbury, 2020) and, with David Lowe, *Remembering the Cold War: Global Contest and National Stories* (Routledge, 2013).

Krishan Kumar is University Professor and William R. Kenan, Jr., Professor of Sociology at the University of Virginia. He was previously Professor of Social and Political Thought at the University of Kent at Canterbury. He has also been a Visiting Professor at the Central European University. Among his publications are *1989: Revolutionary Ideas and Ideals* (2001), *The Making of English National Identity* (2003), *Visions of Empires: How Five Imperial Regimes Shaped the World* (2017), and *Empires: A Historical and Political Sociology* (2021).

Simon Lewis is Associate Professor in East and Central European Cultural History at the University of Bremen. He researches memory, postcolonialism and multiculturalism in East-Central Europe, and is co-chair of the Polish Memory Studies Group, a regional group of the Memory Studies Association. He is the author of *Belarus – Alternative Visions: Nation, Memory and Cosmopolitanism* (2019) and co-author of *Remembering Katyn* (2012). Forthcoming books include *Multicultural*

Commonwealth: Poland-Lithuania and its Afterlives (co-edited with Stanley Bill).

David Lowe is Chair in Contemporary History at the School of Humanities and Social Sciences, Deakin University. David's recent publications include *Remembering Independence* (with Carola Lentz; Routledge, 2018). His research centres on cultural aspects of the history of international relations, including Australia's role in the world; and on remembering the legacies of modern wars and empires in comparative contexts. He is currently researching the history of postwar foreign aid, including the Colombo Plan and Australia's foreign aid program.

Katharine McGregor is a historian of modern Indonesia, based at the School of Historical and Philosophical Studies, University of Melbourne. The focus of her research to date has been on the Indonesian military, the Cold War, colonial violence, memory and activism, Indonesian women, and the Japanese occupation. She has published four books and is currently completing a monograph on transnational activism for Indonesian comfort women, a major outcome of an Australian Research Council Future Fellowship. She recently commenced work on a new Australian Research Council funded project on memory activism around colonial violence, and is collaborating on a new project entitled *Rethinking Colonial History: Indonesia*. She is currently President of the Asian Studies Association of Australia and Deputy Associate Dean International Indonesia for the Faculty of Arts at the University of Melbourne.

A. Dirk Moses is the Anne and Bernard Spitzer Professor of Political Science at the City College of New York. He is the author of *The Problems of Genocide: Permanent Security and the Language of Transgression* (Cambridge, 2021) and senior editor of the Journal of Genocide Research.

Jeffrey K. Olick is William R. Kenan, Jr. Professor of Sociology and History at the University of Virginia (USA) and co-president of the Memory Studies Association. Recent publications include the six-volume *A Cultural History of Memory*, co-edited with Stefan Berger (Bloomsbury, 2020) and a special issue of the journal *Memory Studies* (2021), co-edited with Hanna Teichler, on "Memory and Crisis".

Malgorzata Pakier is Head of the Academic Department at the European Network Remembrance and Solidarity. She is the author of *The Construction of European Holocaust Memory: German and Polish Cinema* *after 1989* (2013) and co-editor of *A European Memory? Contested Histories and Politics of Remembrance* (2010, 2012) and *Memory and Change in Europe. Eastern Perspectives* (2015). Her academic interests include: Europeanization of memory, Holocaust representation, social/cultural memory, museum studies, and film.

Laura Pozzi is Research Assistant Professor at the Faculty of History, University of Warsaw. She is interested in entanglements between global history, colonialism, and memory politics in the People's Republic of China. She studies the impact of postcolonial and decolonial theories on the changing representations of Chinese history in museums and heritage sites at a global level.

Ann Rigney is Professor of Comparative Literature at Utrecht University. She has published widely on theories of cultural memory and on memory cultures in Europe since 1800, including *The Afterlives of Walter Scott: Memory on the Move* (Oxford UP, 2012) and, co-edited with Chiara De Cesari, *Transnational Memory: Circulation, Articulation, Scales* (De Gruyter, 2014). She currently directs an ERC-funded project called *Remembering Activism: The Cultural Memory of Protest in Europe* (ReAct).

Ksenia Robbe is a Senior lecturer in European Culture and Literature at the University of Groningen. She works at the interfaces of postcolonial and postsocialist studies with a focus on memory and time, and writes about contemporary Russian/Russophone and Southern African cultural production. She is the author of *Conversations of Motherhood: South African Women's Writing Across Traditions* (2015) and co-editor of *Post-Soviet Nostalgia* (2019) and *(Un)timely Crises: Chronotopes and Critique* (2021).

Hanna Teichler is a research associate in the department of Anglophone Literatures and Cultures at Goethe University, Frankfurt. She holds a PhD from the department of Anglophone Literatures and Cultures, Goethe University Frankfurt, and a M.A. degree in English, French and Portuguese philology. Her first monograph *Carnivalizing Reconciliation* came out with Berghahn in 2021. Hanna is the co-editor (with Rebekah Vince) of the book series Mobilizing Memories and of the Handbook Series in Memory Studies (both Brill). She was a member of the Memory Studies Association Executive Committee (2018–2022) and co-directs (with Astrid Erll) the Frankfurt Memory Studies Platform.

Joanna Wawrzyniak is Associate Professor of Sociology and Director of the Center for Research on Social Memory at the University of Warsaw. Specializing on East-Central European memory processes, her current projects include memories of socialism, neoliberal transformation, and deindustrialization processes in Poland. She also participates in collaborative research on cultural heritage and memory processes in Western Europe and East and South Asia. She is co-author of the award-winning *Cuts: An Oral History of the Transformation* (2020, in Polish) and *The Enemy on Display* (2015), and the author of *Veterans, Victims and Memory* (2015).

List of Figures

The Cold War and Regions of Memory

Partisan History and the East European Region of Memory

China, the Maritime Silk Road, and the Memory of Colonialism in the Asia Region

Introduction: Regions of Memory in Theory

Simon Lewis and Joanna Wawrzyniak

Despite many voices that have predicted the decline of memory studies after its national boom in the 1980s and 1990s, the next twenty years opened up its next, cosmopolitan chapter. The so-called third wave of memory studies brought a variety of new concepts to account for the unprecedented transnationalization of research in the humanities and social science in general, and in memory studies in particular, such as "multidirectional memory" (Rothberg 2009), "travelling memory" (Erll 2011), "memory unbound" (Bond, Craps, and Vermeulen 2017), "transnational memory space" (Wüstenberg and Sierp 2020), and "mnemonic solidarity" (Lim and Rosenhaft 2021). This book sets to introduce one more notion into the family of memory studies' concepts, that of "regions of memory". Its principal aim is to challenge further the dichotomy of national vs. cosmopolitan memory that was introduced by the third wave

S. Lewis
University of Bremen, Bremen, Germany
e-mail: lewis@uni-bremen.de

J. Wawrzyniak (✉)
Faculty of Sociology, Center for Research on Social Memory
University of Warsaw, Warszawa, Poland
e-mail: wawrzyniakj@is.uw.edu.pl

S. Lewis et al. (eds.), *Regions of Memory*, Palgrave Macmillan Memory Studies, https://doi.org/10.1007/978-3-030-93705-8_1

of memory studies, and thereby to contribute towards a more differentiated study of global memory processes.

The volume examines the scale of collective memory that transcends the national level, yet is narrower than the global. This theoretical intervention enables diverse comparisons in memory studies, putting the most established paradigms of (West) European and (North) American memory in corresponding view alongside other regions such as East Asia, Eastern Europe, and Southern Africa. Memory at the regional level has been studied on different continents, but these projects often tend to be inward-looking compendia of studies in and between nations within those regions (e.g. Allier-Montaño and Crenzel 2015; Kim and Schwartz 2010; Makdisi and Silverstein 2006; Waterson and Woon 2012; Pakier and Wawrzyniak 2015; Subotić 2019). The book builds on such area-specific scholarship in two ways. First, it enables studying a variety of global memory regions in parallel. Second, however, it seeks to highlight processes that transcend geographical boundaries to reveal lesser-known vectors and mechanisms of memory travel, such as across Cold War battle lines or between Indonesia and western Europe. Regions of memory, as this volume shows, are not just geographical territories, but conceptual spaces, united by particular histories and cultures as well as by geographical elements.

More specifically, the volume deals with regions as spaces for memory dynamics in a multiplicity of inter- and transnational situations, and the region-based studies we present in this volume emphasize subjectivities, agencies and internal dynamics within regions, and their entanglements, rather than projections of representations that objectivize space. The concept of memory regions allows us also to interpret the (frequently though not necessarily antagonistic) points of contact between larger and smaller memory formations, e.g.: between the official memory projects of supranational alliances and local memory; between cosmopolitan and vernacular memory; or between the transnational flow and local reception of memory narratives.

Thinking Regionally About Memory

The idea behind "regions of memory" originated at the annual conferences series *Genealogies of Memory in Eastern and Central Europe*, organized by the European Network Remembrance and Solidarity since 2011. The first conferences were shaped by memory debates proliferating after the accession of the East-Central European states to the European Union

(EU) in 2004. They focused on differences and conflicts between the cosmopolitan Western European mode of remembrance of the Second World War dominated by the Holocaust and universalist human rights discourse (e.g. Levy and Sznaider 2006; Bull and Hansen 2016) and the new member states' historical narratives centered around the Gulag experience, national victimhood, wartime heroism, criminalization of communism, and reluctance to admit the local populations' compliance with both Nazi and communist atrocities (e.g. Blaive et al. 2013; Mälksoo 2010; Neumayer 2019; Subotić 2019). Also, there were important discursive similarities between the "eastern" EU states in the processes that shaped their post-accession memory policies, and memory negotiations within the bloc. In this context, we tried to capture both the differences with Western Europe and similarities across East-Central Europe by conceptualizing these spaces as regions of memory.

At the time, Jeffrey K. Olick proposed to define "regions of memory" as "discursive arenas above the level of the nation state but not fully universal" (Olick 2015, xii). The notion was intended to challenge both national and cosmopolitan approaches to memory which dominated the scholarship of the time. Together with Małgorzata Pakier, we also called for a further historization of memory research, as we thought that it was the experience of the twentieth century's mass violence—whose scale in Eastern Europe was incomparable to that in Western Europe—that contributed to particular constellations of images and narratives on the past in this region (Wawrzyniak and Pakier 2013). Finally, at the Genealogies of Memory conferences, we started to compare the configurations of East-Central European memories with those of other regions of the world, such as the Caucasus, East Asia, Latin America and West Africa, whose historical legacies also include experiences of mass violence, fragmentation and diversity. Sometimes we were struck by important similarities, like in the memory dynamics of the Korean-Japanese-Chinese and Polish-German-Russian triangles, in which war, mass killing and territorial conquest continue to divide remembrance across country lines. Most significantly, however, we were enriched by criticisms of our initial approach's geographical essentialism. Regions of memory, several speakers argued, can be formed also across continents or social spectra, without gaining global significance. Taking those discussions into account, we decided that the way to approach regions of memory is to use this notion as what American sociologist Herbert Blumer once called a sensitizing concept. According to Blumer, sensitizing concepts suggest "directions along which to look", in contrast to definitive concepts which provide "prescriptions of what to see" (Blumer 1954, 7). Let us comment on two of the directions that inform this volume.

Whilst not mutually exclusive, the two main directions of this volume differ in their departure points. The first stems from the classical idea of regional space. In this case, the concept of regions of memory is an invitation to revisit area studies, the Cold War intellectual product of American universities with their regional, geographical foci; or as other want to see area studies—an effect of longer term Western World's domination over the global "rest" in the colonization and decolonization processes. Martin W. Lewis and Kären E. Wigen argued in an influential work a quarter of a century ago against "metageographical mythology" that divided the world into implicitly hierarchized units: "continents, nation states, and supracontinental blocks, such as East and West" (Lewis and Wigen 1997, 6, 10). At the same time, they declined to discard spatial supranational constellations altogether, arguing that "[t]he project of metageographical reform must face the challenge of coming up with multiple crosscutting and overlapping regionalisms for different purposes" (ibid., 199). With the idea of "regions of memory" we take up this challenge, adapting it for the specific realm of memory studies.

Despite its political ambiguity, the traditions of area studies resulted in the production of important expertise that transcended national and disciplinary borders, and contributed to in-depth understanding (or construction, as the opponents of this approach would argue) of entangled regional histories, cultural and political processes in various localities from the Latin America, and different parts of Africa, Asia, Australia and New Zealand. Memory studies can draw on and contribute to such area studies expertise without entirely falling into its initial traps, such as modernization theory (putting the "West" as the blueprint for the "rest") or orientalism (treating the "rest" as different from the "West"). The processes of "inventing" or "imagining" regions such as Eastern Europe (Wolff 1994) or the Balkans (Todorova 1997), of making up their mythologies and traditions by both outsiders and local elites, are exactly what memory studies scholars are trained to study. The regions of memory concept, even when referring to geographical spaces, is sensitive of their discursive constructions. As Diana Mishkova and Balazs Trenscényi put it in their comment on the role of "regions" for historiography:

> crucial to this understanding of space is not so much its material morphology as the premises of its social production, its ideological underpinnings, as well as the various forms of interpretation and representation that it embodies. (Mishkova and Trenscényi 2017, 11)

Regions of memory are thus to be studied critically and heuristically, and emphatically not as reifications of geographical space. Even when understood spatially, "regions of memory" are of course not intended as large areas in which everyone shares the same memories—analogously to national memoryscapes, which are likewise never monolithic. There are diverse historical events remembered with varied significance across the geographical space; and the same events are often given different or conflicting meanings. However, their memories are in one or other way discursively connected to the place in which they happened. They might form supra- or transnational constellations of representations of the past within or referring to the particular regional space. They may share specific regional carriers, forms, agents, sites, or nodes of memory.

To come back to our initial example, the extensive memory research undertaken in East-Central Europe in recent decades has uncovered different constellations of collective memory unfolding to this region. The legacies of its ethnic, linguistic, cultural, and religious heterogeneity, early modern and modern imperial histories, nation-building processes, communist experience, border shifts, migration waves, as well as links to diasporas all over the world, formed varied, inter- and transnational constellations of memory. In other words, "East-Central Europe" is not a single region of memory shaped by a specific transnational history; rather, the space commonly referred to using this term contains multiple overlapping and intersecting memory regions that together form uneven concentrations of mnemonic interaction.

To give just a few illustrations of recent research, Natalia Nowakowska (2019) studies memories of the Jagiellonian dynasty—who ruled in Poland-Lithuania, Bohemia, and Hungary in the fifteenth and sixteenth centuries—and discovers a regional framework of remembrance of this family of monarchs, in the sense of a mnemonic system that evolves and changes over centuries and connects the lands of the former Jagiellonian dynasty. In the present moment, Nowakowska concludes, one can speak of nationalization of the Jagiellonians in particular countries on the one hand, while on the other hand the roots of their representations, and the histories of the changes in those representations, are culturally and historically entangled. The same can be said about myriads of other representations originating in the early modern or modern era in the region, layered by and framed by successive generations and changing cultural, social, and political circumstances.

To present an example of twentieth century memories, a team of authors led by Alexander Etkind and Rory Finnin (2012) trace the circulation of cultural representations of the 1940 Katyn massacre, in which more than 22,000 Polish soldiers were murdered by the Soviet NKVD, across Poland, Ukraine, Belarus, Russia and Baltic States. Again, the researchers observe both the movement of images and interpretations across state borders and their anchoring in specific national cultural setting—as well as important remembrance roles outside the region, for instance the impact of political decisions in the USA and the UK, often influenced by Polish diaspora communities.

Finally, one of the largest bilateral multi-volume Polish-German history projects of recent years focuses on the Polish-German *lieux de mémoire*, showing how trans-local symbols and patterns of meaning are formed between these two cultures and others in the region (Traba and Hahn 2012–15; 2015–16). What all these examples—and numerous others—have in common is that they show such a movement of cultural memories which cannot be seized in national containers. Their relevance crosses national borders, however is unlikely to attain global significance, remaining within spatial, regional boundaries.

The other way to conceptualize regions of memory is to draw on the scholarship that postulates to concentrate on the travels, movements, and circulations of memory (Erll 2011; De Cesari and Rigney 2014). Take for instance, the seminal concept of "multidirectional memory" advanced by Michael Rothberg (2009), who focuses on how the memory of the Holocaust was shaped by, confronted by, and contributed to the articulation of histories of other cases of extreme violence in various local settings, including in North America, the Middle East, Eastern Europe, and the Balkans. Rothberg observes that in each of these cases, the memory of the Holocaust plays a different role and has different meanings. In the processes of cross-referencing with other memories, it contributes to a (re)construction of subjects, spaces, and sites of memory.

More generally, the second direction takes as its starting point not a predeterminate geographical space—and a question what constellations of memory nodes, agents, carriers, forms, events, sites, or practices, add to its specificity—but the *movement* of some of these across space; though again, within (loosely) bounded limits depending on the subject being remembered. Importantly for several chapters in this volume, following the movements of remembrance helps to account on the one hand for regionalization of seemingly cosmopolitan themes (e.g. genocide or the Cold War), and on the other hand to identify regions of memory that are beyond the geographical focus of classical area studies. Considering these two directions together, we have realized that the more open ended our understanding of regions of memory is, the more heuristic value it has. Regions of memory can be historically conditioned patterns of cultural similarity, institutional projects for political integration, or imagined constellations of mnemonic solidarity. They can be all of these things at the same time, or any number of other configurations of transnational memory flow—in geopolitical, digital, or artistic space. Regions of memory can be geographically contiguous or dispersed, and have uneven distribution of mnemonic intensity; to return to an earlier example, the memory of the Katyn massacre is not as powerful in Belarus or Latvia as it is in Poland, and also has historical roots in diaspora communities in the USA, UK, and other countries.

Regions of memory are not the "areas" of "area studies", nor are they bound to theoretical constructions such as postcolonialism. Rather, the study of memory regions allows us to reach across different disciplinary and geographically-specialized traditions and enter into multiple forms of dialogue; for example, to territorialize postcolonial discourses by specifying the regions they imagine (see e.g. chapters by McGregor, Kumar, Moses, Teichler and Robbe), or to challenge established spatializations of memory by considering new material, such as the Cold War (Joel and Lowe), the One Belt, One Road initiative (Pozzi), or the digitalization of memory through a resource like Wikipedia (Rigney). The concept of memory regions should also be empowering rather than essentializing, as it allows for the locally situated voices within cultures to be heard and understood as constitutive co-creators of mnemonic transculturation—as all the chapters that follow illustrate, in different ways. Regions of memory, in other words, do not need to be thematically or geographically bounded. The borders of memory regions define themselves, depending on what is being remembered and by whom.

Structure of the Volume

The volume is divided into three sections, illustrating three broad categories through which regions of memory may be analyzed: **historical**, **political**, and **cultural** regions of memory. These categories are far from exhaustive, and not mutually exclusive. In considering "historical" regions of memory we take as our point of departure the well-established fact that memory cultures are themselves products of history. As Olick (1999, 383) puts it, remembrance involves "not only the history being commemorated but also the accumulated succession of commemorations, as well as what has occurred between those powerful moments." Societies that have historically strong ties, or that have been affected by similar historical events, are more than likely to cultivate and exchange memories about the shared past, setting off further path dependencies and chains of dialogue. The paradigmatic example here is the Holocaust, whose "Americanization" and "Europeanization" have made it the primary candidate for a "globalized" status (e.g. Levy and Sznaider 2002, 2006; Assmann 2010). Yet the Holocaust is clearly not remembered in similar ways or to a similar extent everywhere, with phenomena such as migration complicating the picture even at the European core (Rothberg and Yildiz 2011). Holocaust recognition may have been a "contemporary European entry ticket" for many Eastern European countries in the early-to-mid-2000s before EU accession (Judt 2005, 803), but a decade and a half later, amidst a rise of nativist right-wing populist governments in the region that have downplayed its significance, it is clear that supranational specificities remain.

The three chapters in the first section of the book explore such historically conditioned memory regions, but they do not only show how transnational historical events create singular regions of memory. Rather, they analyze the interactions and cross-flows between multiple historical legacies and the ways in which different regionalized memory regimes can mutually influence and interrupt each other. **David Lowe and Tony Joel** take the Cold War, a global event (or set of events) with no clear temporal boundaries or even main actors, and show how its remembrance is at once universal and specific: the Cold War is commemorated all over the world, but in different regions it is remembered in different border-crossing constellations. The chapter examines four case studies, two European political projects and two socio-political phenomena in the Asia-Pacific: first, the Visegrád Group comprising of Czechia, Hungary, Poland, and Slovakia; second, the Baltic Initiative and Network incorporating 11 European countries; third, the "nuclear colonialism" whereby world powers used Pacific islands as their testing grounds for atomic warheads; and fourth, the rise of tourism inflected by a Cold War past on the islands of Okinawa (Japan) and Kinmen (Taiwan). The examples are geographically dispersed and thematically diverse, encompassing various levels of state involvement and grassroots activity. Yet in juxtaposition, they show that the Cold War is remembered in distinct memory regions that are tied together by threads of commonality.

Katharine McGregor's chapter also examines the overlapping of different memory regions, and the past being remembered is likewise connected to the Cold War: the 1965–68 mass murder of approximately half a million people by government forces in Indonesia under the guise of an anti-communist purge. McGregor analyzes the intersections between what she terms "human rights regions of memory", whereby she ascribes ethical rather than geographical connotations to the notion of "regions". One region of memory is the human rights discourse centered around the Hague tribunals: the author argues that the convening in 2015 of the International People's Tribunal on Crimes against Humanity in Indonesia 1965 in the Dutch city tapped into the "global memory of human rights injustices". Another thematically defined region is the memory of anti-communism, including a historical commitment of western governments to fight perceived communists in post-war decades; the People's Tribunal also accused the Netherlands, the USA and Australia of complicity in the atrocities. A third region is the postcolonial memory region: in this case, representatives of the Indonesian government countered with arguments that the activists were marginalizing the colonial violence of Dutch imperialism in Indonesia by bringing the tribunal to the Netherlands. The chapter shows skillfully how these different historical legacies—each of them transnational in different ways—intertwine and are brought into contact when the memory of the genocide is activated.

Krishan Kumar's chapter examines the multiple imperial legacies that crystallize in the region of East-Central Europe, creating a region of memory that is defined by its location on the fault-lines of land-based empires: the Holy Roman Empire, the Habsburg Monarchy, the Ottoman Empire, the Hohenzollern dynasty, the Romanov and Soviet realms of expansion, and—briefly but most bloodily—the Nazi empire. He demonstrates the contrasting timbres of remembrance that result from the entanglement of imperial histories, from nostalgia for Habsburg rule to politicized recriminations over the divisive legacies of communism. Imperial histories are by definition transnational, and the memories that circulate in the aftermath of empires likewise resonate at the regional level.

The second section of the book concerns regions of memory as political projects, both as supranational or imperial administrative initiatives to create a common realm of remembrance, and as conglomerations of political fields that converge to create regionalized patterns. The paradigmatic case of the former is, of course, the European Union, which through its decades of existence in various guises has gradually come to espouse a more or less unified—if, needless to say, still contested—memory policy through legislation, political declarations, the founding of a pan-European memory museum in the form of the House of European History in Brussels, and other measures (e.g. Sierp 2014; Koposov 2017; Verovšek 2020). Another example is the post-Soviet realm, where the exigencies of post-Soviet transition and present-day Russian expansionism have combined to create a territory where old myths of Soviet victory in the Second World War (known here as the Great Patriotic War, 1941–45) are valorized by many—especially in Russia and Belarus and some parts of Ukraine—and rejected by an increasing number of societies in the former peripheral republics (Fedor et al. 2017). These established cases show that political memory regions can work to integrate societies across national borders, but also that political systems play a major role in how successful such endeavors may be. As a union of participatory democracies, the EU's striving for common remembrance is influenced to a greater extent by transnational grass-roots activity and mnemonic pluralism. Memory in the post-Soviet realm, in contrast, is shaped above all by autocratic regimes and their official ideologies, as well as oppositional narratives that emerge to debunk those ideologies. The tensions that result in each type of memory region are a function of their political structures.

Laura Pozzi's chapter considers a different supranational project: the "One Belt, One Road" initiative launched by the Chinese government in 2013. This foreign policy program has typically been seen as a politico-economic phenomenon, but Pozzi shows that it also has an effect on narratives of remembrance. She analyzes three museums: the Shanghai History Museum/Shanghai Revolution Museum, the Hong Kong Museum of History, and the Galle National Museum in Sri Lanka. These three institutions are examined on a continuum: in Shanghai, the Chinese government exerts full control over the narrative, while in the Hong Kong Special Administrative Area the influence of the mainland is strong though not total, and political pressure is on the rise; the Sri Lankan case, on the other hand, is located on the periphery of Chinese influence, and the creation of a gallery dedicated to the "Sri Lanka-China Historical Relationship" reveals both an attempt to construct a Sinocentric pan-Asian memory, and also the limitations thereof; the display is fairly marginal within the museum building. Chinese hegemony in Asia is, therefore, a region of memory in the making. How successful this initiative can be is a subject for further research.

A. Dirk Moses, in turn, explores the Eastern European memory region not as a supranational institutional effort (like the Visegrád Group analyzed by Joel and Lowe), but as a realm of "partisan histories". This term captures both the "highly partial" mnemonic narratives that circulate in political discourse in several East European states and the historical phenomenon of "partisans", i.e. mid-twentieth century freedom fighters who pursued political goals by military means and who remain key objects of commemoration by memory activists today. Moses demonstrates that the politicization of memory occurs transnationally when political actors choose in parallel to instrumentalize and nationalize the notion of genocide—paradoxically, a term intended by its inventor Raphael Lemkin as a cosmopolitan idea, never a national one. The proliferation of partisan histories throughout Eastern Europe is of course a result of overlapping historical legacies, including imperialism and epochal violence (as discussed in Kumar's chapter), but importantly, Moses also shows that migration and diaspora play important roles in negotiations over the meanings of the past. Using examples from North America and Israel, he demonstrates that this region of memory extends beyond the confines of continental Eastern Europe.

The third part of the volume considers cultural regions of memory. As is now well established, memories are by nature mediated, with symbols, texts and images playing key roles to "mediate between individuals and, in the process, create communality across both space and time" (Erll and Rigney 2009, 1). The three chapters in this section consider ways in which regions of memory are constituted by writers and readers particular, with one study on Wikipedia as a realm of supranational but bounded memory, and two analyses of African literary constellations that construct memory regions. As **Ann Rigney** points out in the opening theoretical section of her chapter, classical theories of nationalism such as those of Benedict Anderson (1983) envisioned literacy and literary imaginations as preconditions of bounded national communities; however, a similar logic can be applied to examine ways in which contemporary "complex media ecologies" construct larger transnational networks of remembrance and identity.

Rigney builds on her own previous work (De Cesari and Rigney 2014) to elaborate a theory of *articulation*, noting the double meaning of this word in English: both speaking and connecting. The articulations of memory, she argues, are important precisely because speech acts and mediations of language and image carry out the work of "performing connections". Rigney distinguishes four types of articulation: first, *integrative articulation*, the classic example of which is nineteenth-century nation-building, whereby distant poles of the same community are drawn together through redemptive narrative; second, *multidirectional articulation*, whereby seemingly unrelated historical events are drawn together in the same text; third, *hetero-remembrance*, whereby stories are told about communities and peoples that are "other" vis-à-vis the authors and the majority of readers or viewers; and fourth, *translation*, in some ways the opposite of hetero-remembrance in that "it involves importing and making legible narratives that originated elsewhere". She then goes on to apply these concepts to a survey of Wikipedia as a realm of different digital articulations of memory, in which diverse historical topics are covered in vastly different ways—both qualitatively and quantitatively—across languages and societies. The variegated patterns of remembrance clearly transcend national boundaries, but depending on what is being remembered, create differentiated landscapes of regionalized memory.

Ksenia Robbe studies the construction of a southern African arena of memory dialogue by comparatively analyzing books of diverse genres and a feature film, from South Africa, Zimbabwe, and Namibia. Whilst the three countries experienced different colonialisms—Dutch, British, German, and Apartheid—their histories of both colonization and decolonization were "deeply entangled". Robbe thus discovers a commonality of priorities in the "alternative memories of decolonization" in all three countries, which each provide critical counternarratives to the nationalist martyr stories of official memory. She compares two separate generations of cultural figures in order to argue for a shift in the modality of remembering across the region: whereas in the late 1990s and early 2000s, a predominant preoccupation of literature was "witnessing", more recent work has focused on the postmemory of later generations (Hirsch 2008) and the restoration of decolonial subjectivities in the colonial past—Robbe coins the term *activist genealogies* to describe this trend. She analyzes three novels of the earlier generation as representative examples of the "witnessing" of colonial crimes: Kaleni Hiyalwa's *Meekulu's Children* (Namibia, 2000), Yvonne Vera's *Stone Virgins* (Zimbabwe, 2002), and Zoë Wicomb's *David's Story* (South Africa, 2000). "Activist genealogies" are found in Koleka Putuma's poetry volume *Collective Amnesia* (South Africa, 2017), the non-fiction prose work *These Bones Will Rise Again* by Panashe Chigumadzi (Zimbabwe, 2018), and the feature film *The Unseen* by Perivi Katjavivi (Namibia, 2016). The chapter demonstrates the heterogeneity and polyphony of the southern African memory region, with a plethora of distinct voices and diverse storylines, but also the clear and coherent patterns of memory that can be observed in all three countries.

Hanna Teichler, in contrast, argues for the transcontinental entanglement of African memories as a consequence of the global history of slavery and indentured labor. She studies two anglophone Kenyan novels, M.G. Vassanji's *The In-between World of Vikram Lall* (2003) and Yvonne Adhiambo Owuor's *Dust* (2014), which both reach across oceans to remember, respectively, the legacies of Asian chattel labor in East Africa and the transatlantic slave trade between the African continent and South America. Vassanji's novel centers on its eponymous main character, the descendant of a Punjab laborer, as he grows up in British-ruled East Africa and matures in independent Kenya. Owuor's novel sees its protagonist travel to Brazil, where the legacies of slavery haunt her experiences and the novel as a whole. Analysis of these two works shows that postcolonial regions of memory are transoceanic, exploding any perceived geocultural divide between the mnemonic spaces of the Indian and Atlantic Oceans. Teichler also views them as cosmopolitanizing forces in modern Kenyan national identity, forging outward gazes on the present and future through their multi-layered and multidirectional treatments of the past.

Clearly and inevitably, the volume has thematic and geographical gaps. Critics may argue that we could have included chapters with a focus on Middle Eastern or Latin American memory, religious regions of memory, musical genres creating memory regions, megalopoli or cultural capitals as the centers of regions, or many other possible and/or real transnational constellations of commemoration. Nonetheless, this book is intended as a conversation starter on the phenomenon of regions of memory, not as an exhaustive catalogue of the possible configurations thereof. What we believe the volume does show is the viability of regions of memory as a category of analysis. Memory regions enable a transcendence of the "container" approach to national memory and analysis of "how memory travels" (Erll 2011) whilst maintaining a spatial focus on *where* and *how intensively* memories move, and what they connect. In adding a sensitivity to the boundedness and cultural specificity of transnational remembrance, the concept serves to deconstruct the universalizing abstractness of supposed "cosmopolitan" ideas of global memory—which may paradoxically conceal unconscious Eurocentric biases.

References

Allier-Montaño, Eugenia and Emilio Crenzel, eds. 2015. *The Struggle for Memory in Latin America: Recent History and Political Violence*. Basingstoke: Palgrave Macmillan.

Anderson, Benedict. 1983. *Imagined Communities: Reflections on the Origins and Spread of Nationalism*. London: Verso.

Assmann, Aleida. 2010. The Holocaust—a Global Memory? Extensions and Limits of a New Memory Community. In *Memory in a Global Age: Discourses, Practices and Trajectories*, ed. Aleida Assmann and Sebastian Conrad, 97–117. New York: Palgrave Macmillan.

Blaive, Muriel, Christian Gerbel, and Thomas Linderberger, eds. 2013. *Clashes in European Memory: The Case of Communist Repression and the Holocaust*. Innsbruck: StudienVerlag.

Blumer, Herbert. 1954. What is Wrong with Social Theory? *American Sociological Review* 19, no. 1: 3–10.

Bond, Lucy, Stef Craps and Pieter Vermeulen, eds. 2017. *Memory Unbound: Tracing the Dynamics of Memory*. New York: Berghahn.

Bull, Anna Cento and Hans Lauge Hansen. 2016. On agonistic memory. *Memory Studies* 9, no. 4: 390–404.

De Cesari, Chiara and Ann Rigney. 2014. Introduction. In *Transnational Memory: Circulation, Articulation, Scales*, ed. Chiara De Cesari and Ann Rigney, 1–25. Berlin. De Gruyter.

Erll, Astrid. 2011. Travelling Memory. *Parallax* 17, no. 4: 4–18.

Erll, Astrid and Ann Rigney. 2009. Introduction: Cultural Memory and its Dynamics. In *Mediation, Remediation, and the Dynamics of Cultural Memory*, ed. Astrid Erll and Ann Rigney, 1–11. Berlin: De Gruyter.

Etkind, Alexander, Rory Finnin, et al. 2012. *Remembering Katyn*. Cambridge: Polity.

Fedor, Julie, Markku Kangaspuro, Jussi Lassila, Tatiana Zhurzhenko, eds. 2017. *War and Memory in Russia, Ukraine and Belarus*. New York: Palgrave Macmillan.

Hirsch, Marianne. 2008. The Generation of Postmemory. *Poetics Today*, 29, no. 1: 103–28.

Judt, Tony. 2005. *Postwar: A History of Europe Since 1945*. New York: Penguin.

Kim, Mikyoung and Barry Schwartz, eds. 2010. *Northeast Asia's Difficult Past: Essays in Collective Memory.* Basingstoke: Palgrave Macmillan.

Koposov, Nikolay. 2017. *Memory Laws, Memory Wars: The Politics of the Past in Europe and Russia.* Cambridge: Cambridge University Press.

Levy, Daniel and Natan Sznaider. 2002. Memory Unbound: The Holocaust and the Formation of Cosmopolitan Memory. *European Journal of Social Theory* 5: 87–106.

Levy, Daniel and Natan Sznaider. 2006. *The Holocaust and Memory in the Global Age.* Trans. Assenka Oksiloff. Philadelphia: Temple University Press.

Lewis, Martin W. and Kären E. Wigen. 1997. *The Myth of Continents A Critique of Metageography.* Berkeley: University of California Press.

Lim, Jie-Hyun and Eve Rosenhaft. 2021. *Mnemonic Solidarity: Global Interventions.* Cham: Palgrave Macmillan.

Makdisi, Ussama Samir and Paul A Silverstein, eds. 2006. *Memory and Violence in the Middle East and North Africa.* Bloomington: Indiana University Press.

Mälksoo, Maria. 2010. *The Politics of Becoming European: A Study of Polish and Baltic Post-Cold War Security Imaginaries.* London and New York: Routledge.

Mishkova, Diana and Balázs Trenscényi. 2017. Introduction. In *European Regions and Boundaries: A Conceptual History,* ed. Diana Mishkova and Balázs Trenscényi, 10–25. New York: Berghahn Books.

Neumayer, Laure. 2019. *The Criminalisation of Communism in the European Political Space after the Cold War.* London and New York: Routledge.

Nowakowska, Natalia, ed. 2019. *Remembering the Jagiellonians.* Abingdon: Routledge.

Olick, Jeffrey K. 2015. Foreword. In *Memory and Change in Europe: Eastern Perspectives*, ed. Małgorzata Pakier and Joanna Wawrzyniak, ix-xii. New York: Berghahn Books

———. 1999. Genre Memories and Memory Genres: A Dialogical Analysis of May 8, 1945 Commemorations in the Federal Republic of Germany. *American Sociological Review* 64, no. 3: 381–402.

Pakier, Małgorzata and Joanna Wawrzyniak, eds. 2015. *Memory and Change in Europe: Eastern Perspectives*. New York: Berghahn Books.

Rothberg, Michael. 2009. *Multidirectional Memory: Remembering he Holocaust in the Age of Decolonization.* Stanford: Stanford University Press.

Rothberg, Michael and Yasemin Yildiz. 2011. Memory Citizenship: Migrant Archives of Holocaust Remembrance in Contemporary Germany. *Parallax* 17, no. 4: 32–48.

Sierp, Aline. 2014. *History, Memory, and Trans-European Identity: Unifying Divisions.* New York: Routledge.

Subotić, Jelena. 2019. *Yellow Star, Red Star: Holocaust Remembrance after Communism*. Ithaca and London: Cornell University Press.

Todorova, Maria. 1997. *Imagining the Balkans.* Oxford: Oxford University Press.

Traba, Robert and Hans Henning Hahn, eds. 2012–15. *Deutsch-polnische Erinnerungsorte*. 5 vols. Paderborn: Ferdinand Schoningh.

———. 2015–16. *Polsko-niemieckie miescja pamięci*. 4 vols. Warszawa: Scholar.

Waterson, Roxana and Kwok Kian Woon, eds. 2012. *Contestations of Memory in Southeast Asia*. Singapore: NUS Press.

Wawrzyniak, Joanna and Małgorzata Pakier. 2013. Memory Studies in Eastern Europe: Key Issues and Future Perspectives. *Polish Sociological Review* 183, no. 3: 257–80.

Wolff, Larry. 1994. *Inventing Eastern Europe: The Map of Civilization on the Mind of the Enlightenment*. Stanford: Stanford University Press.

Wüstenberg, Jenny and Aline Sierp, eds. 2020. *Agency in Transnational Memory Politics*. New York: Berghahn Books.

Verovšek, Peter. 2020. *Memory and the Future of Europe*. Manchester: Manchester University Press.

PART I

Historical Regions of Memory

The Cold War and Regions of Memory

David Lowe and Tony Joel

The Cold War can be broadly defined as an era in which most of the developed world was divided into polarized blocs during the second half of the twentieth century, with the two global superpowers, the Soviet Union and the United States, leading the "East" and the "West" respectively. Sharp lines of division were carved by politico-ideological differences concerning communism on the one hand and democratic capitalism on the other hand. Military tensions always simmered, and the threat of nuclear warfare seemed real and constant. Yet, local skirmishes, isolated acts of violence, and several notable proxy wars notwithstanding, the overarching conflict never reached boiling point—hence the use of the term "cold". Unlike conventional or "hot" wars that begin and end with specific events, there is neither a universally accepted starting date nor a unanimously recognized conclusion for the Cold War (beyond general agreement that it spanned from the 1940s up to the late 1980s or early 1990s around the time of the fall of the Berlin Wall, the collapse of eastern European communism, and the dissolution of the Soviet Union). In any case, the Cold War shaped international relations on a daily basis and fashioned world

D. Lowe (✉) • T. Joel
Faculty of Arts and Education, Deakin University, Waurn Ponds, VIC, Australia
e-mail: david.lowe@deakin.edu.au; tony.joel@deakin.edu.au

S. Lewis et al. (eds.), *Regions of Memory*, Palgrave Macmillan Memory Studies, https://doi.org/10.1007/978-3-030-93705-8_2

history for roughly half a century, and its effects continue to be felt around the globe today—not least when it comes to issues surrounding political legacies and (re-)defining national identities. The era readily lends itself to museum exhibits, filmic representation and popular culture, memorial days and anniversaries, monument building or other forms of memorialization, and also has left behind visible traces of a built environment through historic sites, period architecture, and other structures now valued as cultural heritage worthy of preservation. Much of this activity is undertaken as a nationalistic endeavor, whether it be a top-down approach with state backing, or more organic bottom-up ventures inspired by social agency. Accordingly, as we have explored previously, when it comes to remembering the Cold War, the issue is often framed as a "global conflict" told through "national stories" (Lowe and Joel 2014). But what about the mnemonic spaces in between? In other words, what can be said of regions where Cold War memories transcend national boundaries but still remain localized, rather than universal or even continental?

This chapter presents an exploratory study of the notion of Cold War memory regions by surveying four case studies. It commences with two European examples: first, the Visegrád Group (or V4), comprising of Czechia, Hungary, Poland, and Slovakia; second, the Baltic Initiative and Network, incorporating eleven nations spread across western, central, and eastern Europe. It then examines two cases from the Asia-Pacific: the "nuclear colonialism" that occurred in the Pacific as world powers used others' backyards as their testing grounds; and the rise of tourism inflected by a Cold War past in two Cold War islands, Japan's Okinawa and Taiwan's Kinmen off the coast of mainland China. In all four instances, the various nationalities that form a region are linked through what can be termed as either "shared" or "common" memory. Here, the term "shared memory" relates to a group or groups of individuals (and their ancestors) who experienced and now remember the *same* event, whereas "common memory" denotes when otherwise unrelated groups (and even erstwhile opponents) join together in remembering *cognate* experiences (Lowe and Joel 2014, 14; Ashplant et al. 2000, 18–19). Some of these examples involve formalized collaboration within a region, whereas others comprise of a looser collection of similar yet separate cases. Is "regional" memory of the Cold War motivated by tourism, cultural heritage, history, education, or other factors? In particular, victim/survivor status looms large in the politics of memory as attempts are made to use the Cold War era in shaping and reshaping contemporary relationships within these regions and beyond.

Reflecting on these issues, the chapter concludes with some observations about the key commonalities and dissimilarities that can be drawn between the four case studies and what they tell us about remembering this global event at a regional level.

The Politicization of Victimhood

Central Europe is a region, as Slovak historian Roman Holec has commented, which perhaps has "too much history" (quoted in Terenzani 2010). The rise of nationalism in the late modern period may have further intensified the centuries-old trend of major historical developments often serving to divide rather than unite neighboring peoples in the region, but at certain critical junctures there have been concerted efforts made to unify in pursuit of mutually held ambitions. Amid the collapse of eastern European communism and an emerging new world order, in February 1991 Hungarian leader József Antall welcomed his Czechoslovakian and Polish counterparts Václav Havel and Lech Wałęsa to a summit in the small but historically symbolic fortress town Visegrád. Perched on a bend of the Danube north of Budapest, it was here in 1335—when Visegrád castle served as the royal seat—that Hungarian king Charles I hosted a congress with the Bohemian king John of Luxembourg and the Polish king Casimir III for the similar purpose of advancing mutual cooperation between their three kingdoms and securing peace in central Europe. What primarily motivated these states to form a new alliance in the 1990s was creating a united front to help one another expedite their common goal of western integration—characterized by NATO and European Union (EU) membership—while simultaneously setting about to eliminate "the remnants of the communist bloc" within their region (Visegrád Group n.d.). Whereas their collective focus naturally fixated on future political goals, as evidenced by the choice of Visegrád as the initial meeting place there also was a strong desire from the outset to recognize that within central Europe's congested and complicated history, Czechoslovakians, Hungarians, and Poles shared a common past, too, including most immediately during the Cold War era.

The Visegrád Declaration of 1991 had three signatory states, but following the peaceful separation orchestrated between Prague and Bratislava from 1993 onwards, the Group has become popularly known as "V4" to signify Czechia and Slovakia as discrete members along with Hungary and Poland. Although they have successfully worked together to pursue many common interests, naturally the four states differ considerably in their attitudes toward certain issues (Gyárfášová and Mesežnikov 2016; Klein 2019). Nonetheless, for our purposes here, it is more important to examine how the V4 region was forged out of the Cold War era and subsequently, the politicization of memory has seen its members collectively cast themselves through the prism of victimhood. It has served their purposes well, which should not be surprising given that victimhood has been "vigorously contested" as a political asset since at least the 1990s (Winter and Sivan 1999, 18). In May 1955, Czechoslovakia, Hungary, and Poland became founding members of the Treaty of Friendship, Cooperation, and Mutual Assistance (or so-called Warsaw Pact) shared between the Soviet Union and its communist bloc satellite states in central and eastern Europe. All three states remained committed members for the next 36 years until the Treaty was formally dissolved in July 1991. As such, Czechoslovakia, Hungary, and Poland could be conceived of as cornerstones of the "vanquished" eastern bloc that "lost" the Cold War. They immediately set about very conveniently divorcing themselves from their Warsaw Pact past, however, and instead promoted narratives in which they were long-suffering hapless *victims*, first of Stalinist tyranny and expansionism followed by unwanted and enduring Russian oppression. The V4 members' underlying shared memory of the Cold War, then, is victimization at the hands of Moscow with communism imposed on them at the expense of national self-determination.

Their common memories of such victimization manifests most publicly in remembrance of notable flashpoints that occurred in the various nations. These cognate experiences include, for Hungarians, the uprising of 1956, and for Czechs and Slovaks, the Prague Spring of 1968 (Lowe and Joel 2013, 79–111). While Warsaw was not inflicted by an armed clash on the same scale as the other two capitals, Poles nonetheless can remember popular conflict against state authorities ranging from the Poznań protests of June 1956, to the strikes in Gdynia, Szczecin, and other northern cities in December 1970, through to the Solidarity Movement's emergence from a Gdańsk shipyard in 1980 and the communist government's subsequent declaration of martial law (Judt 2010, 312–13, 589–91, 605–8). The scarring from such exceptional episodes of heightened turmoil and violence could be harnessed under the rubric of the V4 members' wider suffering caused by decades of communist rule, as a means to articulate their post-Cold War era quest for accelerated integration into a broader (read western democratic) Europe. In fact, following the fall of the Berlin Wall,

Czechoslovakia, Hungary, and Poland were so keen to align themselves militarily with their erstwhile foes of the US-led western bloc that they became coalition members of the Gulf War of 1990–91 (at a time when they still remained officially tied to the Warsaw Pact). Just as they had entered into the Warsaw Pact together four decades earlier, the Czechs, Hungarians, and Poles all moved into the NATO orbit in 1999 (with Slovakia following suit five years later). Only Austria, Finland, and Sweden—all nominally neutral states in the Cold War—were invited to join the European Union in the 1990s. But on May 1, 2004, all four Visegrád members were among the eight central and eastern European countries who joined an expanded EU. It had taken the V4 states some thirteen years to realize their common goal of EU accession. It may have been a defining moment in symbolizing their political journey away from the communist past, but their EU existence has not always been smooth sailing and the V4 region's Cold War memories have periodically resurfaced during struggles with Brussels (Fig. 1).

A recent example of V4 leaders appropriating their region's memories of the Cold War era for contemporary politicking unfolded in May 2018. The European Commission (EC) drafted a budget for 2021–27 in which it proposed the introduction of a provocative new mechanism labelled "conditionality", which effectively was designed to see member states' rights to EU funding directly linked to their observance of the rule of law (the Conditionality Regulation eventually was adopted in December 2020). From the outset, the proposal was unmistakably directed against Hungary and Poland, both of whom had previously demonstrated a "blatant disregard of rule of law standards" (Bárd and Śledzińska-Simon 2019, 3). Furthermore, "conditionality" was widely viewed as a thinly veiled form of retaliation against the whole Visegrád Group after all four states had refused to open up their borders in response to the migration crisis of 2015. Three days after the EC released its draft budget calling for "conditionality", EC president Jean-Claude Juncker courted controversy when he visited Trier, the hometown of Karl Marx, on the occasion of the 200th anniversary of the German philosopher's birth. Juncker unveiled a highly divisive 5.5-meter-tall bronze statue of Marx, created by China's leading sculptor Wu Weishan and donated by the Beijing government. Some critics had suggested that Trier should reject the offer or otherwise it could be seen as condoning human rights abuses in China, and the unveiling ceremony was interrupted by protesters (BBC 2018). Striking back within days of being attacked by the "conditionality" innovation, V4 politicians

Fig. 1 The V4 leaders, "REGION OF CREATIVITY", from their 2013/14 summit. (Source: Openverse. https://wordpress.org/openverse/photos/2320c6bd-99a0-4a1d-8750-6caddd4201a9?referrer=creativecommons.org)

seized on the opportunity to criticize Juncker for his apparent insensitivity and called for him to cancel his role in the festivities marking the 200th anniversary. Members of the European Parliament representing Hungary's ruling Fidesz-Christian Democrat alliance called for more respect to be shown to the victims of communism, stating that "Marxist ideology led to the death of tens of millions and ruined the lives of hundreds of millions" (quoted in Visegrád Group 2018). Arguing that "the celebration of its founder is a mockery of their memory", the statement targeting Juncker concluded:

> This is particularly appalling for the citizens of countries which suffered for decades under Communist dictatorships. We oppose the EC head's supportive presence at the celebration, especially since his party family, the European People's Party, is fighting for the EU to condemn the sins of Nazism and Communism.

Undeterred by the V4 backlash, the EC president attended the ceremony in Trier. In his speech, Juncker defended Marx, arguing that his legacy has been unreasonably distorted and thus it is unfair to hold him personally "responsible for all the atrocity his alleged heirs have to answer for" (quoted in Stone 2018). Subsequently, other V4 politicians similarly criticized Juncker for "insulting the memory of millions of victims of communism who were killed, imprisoned or expropriated of their property, in the name of the Marxist doctrine" (Vondra 2018). The ostensibly unrelated issues—remembering Cold War era suffering under communism on the one hand, and contemporary rule of law violations and refusals to accept refugees on the other hand—suddenly became inextricably linked. It created a struggle that Alexandr Vondra, the prominent Czech politician who served time in prison as a dissident shortly before serving as an advisor to Havel immediately following the Velvet Revolution, aptly described in May 2018 as a "Cold War between Visegrád and Brussels" (Vondra 2018). With the EC maintaining pressure on Warsaw and Budapest concerning rule of law infringement proceedings and displeasure over the ongoing migration crisis, Juncker's longstanding sparring partner the Hungarian prime minister Viktor Orbán retaliated in April 2019. Orbán remarked that, after unveiling the statue in Trier the previous year, "Juncker shouldn't be educating us on xenophobia", given that Marx was renowned as the "master of xenophobia" (Hungary Today 2019).

Whereas V4 leaders took umbrage at events marking the 200th anniversary of Marx's birth in May 2018, seven years earlier their region had embraced the centenary of the birth of former US president Ronald Reagan. On June 29, 2011, former secretary of state Condoleeza Rice acted as Nancy Reagan's emissary when a bronze statue (2.2-meter tall) of Reagan was unveiled in Budapest's Freedom Square, adjacent to the US embassy. At the ceremony, Orbán praised Reagan as the leader:

> ...who changed, who renewed, this world and created in it a new world for us in Central Europe—a man who believed in freedom, who believed in the moral strength of freed people and that walls that stand in the way of freedom can be brought down. (Quoted in Birnbaum 2011)

Two days later, Rice travelled to Prague where she again represented the United States at a ceremony to rename the street in front of the US ambassador's residence as *Ulice Ronalda Reagana* in honor of the fortieth president (Johnstone 2011; Embassy of the United States, Prague 2011).

The timing was especially poignant for Czechs, as it coincided with the twentieth anniversary of the Havel administration overseeing the dissolution of the Warsaw Pact when it was signed out of existence in Prague on July 1, 1991 (Havel 2007). Four months later, a project that had been in the works for five years finally came to fruition on November 21, 2011, when a statue of Reagan was unveiled opposite the US embassy in Warsaw. Addressing a smallish crowd full of American and Polish dignitaries, a remarkably modest Wałęsa described himself simply as "a participant" in the events that unfolded across eastern Europe during the 1980s, and questioned whether "today's Poland, Europe and world could look the same without president Reagan" (quoted in Gera 2011). Wałęsa also paid tribute to some of his and Reagan's contemporaries including former British prime minister Margaret Thatcher, France's former president François Mitterrand, and the Polish-born Pope John Paul II as inspirational figures during the final years of Cold War division. He did not include former Soviet leader Mikhail Gorbachev, however, whom Wałęsa has described as a "weak politician" (quoted in Hawley 2009). The very different situations surrounding Reagan's 100th and Marx's 200th anniversaries underline the paradox that ultimately defines the region's Cold War past, some two or three decades later: in 2011, Visegrád countries celebrated the former US president as a defender of freedom and the western democratic leader who gave the region hope it could break free from Soviet domination and the yoke of communism; in 2018, Brussels viewed the V4 states as a problematic bloc still not fully divorced from or "cleansed" of its "eastern" roots, consequently seen as the "other" yet to fully integrate into western thinking (Fig. 2).

Sharing a Divided Past

Geographically speaking, the Baltic Initiative and Network is remarkably close to the Visegrád Group—indeed, the two actually overlap with Poland a member state of both regions. But in many important respects, this is where the similarities end. V4 grew out of the collapse of communism as a state-based endeavor pursuing integration with the democratic west. As the Visegrád countries have collectively reasserted themselves as a new political force in central/eastern Europe, the region has had cause to invoke Cold War memories for contemporary political purposes when dealing with outside pressures or opposition. Conversely, the Baltic Initiative and Network emerged as a grassroots venture. It focused on

Fig. 2 Ronald Reagan monument, a week after its unveiling, Warsaw, November 2011. (Photograph by Tony Joel)

remembering and learning about the Cold War era by bringing together countries that hitherto had been neighbors for centuries, but then suddenly found themselves separated for roughly a half-century and thus experienced the same event from very different perspectives. With an emphasis on preserving historical remnants from the period and supporting educational activities, the Initiative's raison d'être is strengthening "mutual understanding between the countries around the Baltic Sea through an exchange of information on the Cold War period" (Baltic Initiative and Network n.d.).

There is a strong sense of social agency behind the Initiative's origins. It was founded in 2005 by the Danish landscape architect Johannes Bach Rasmussen, inspired by his desire to develop better knowledge and understanding of a world that had been geographically so close and yet for much of his lifetime had remained off limits due to the imaginary "iron curtain" that had descended across the continent. Subsequent revelations about the extent to which western governments in northern Europe had committed themselves to building defense networks, cultivating alliances, and engaging in espionage activities further stimulated Rasmussen's efforts to attract partners to his quest for increasing public awareness of the region's Cold War experiences through imaginative engagement with its physical legacies (Rasmussen, personal correspondence with the authors, 2012). Consultancy work with Danish ministries saw Rasmussen complete a range of tourism and heritage assessments in (neighboring) former communist countries, culminating in his establishment of the Initiative that comprises of eleven member states (Denmark, Estonia, Finland, Germany, Latvia, Lithuania, Poland, Russia, and Sweden plus Iceland and Norway as two exceptions to the Baltic littoral). The Initiative supports, and is supported by, a range of government authorities, institutes, NGOs, museums, and other assorted groups and individuals engaged in activities related to remembering the Cold War era. Its Advisory Group boasts over two dozen honorary figures from across the eleven member states, including former foreign ministers and assorted parliamentarians, ambassadors, retired military commanders, museum directors, cultural heritage practitioners, and emeritus professors among other career backgrounds. Everyone is brought together by the idea that:

> …history should be told from the historically valuable sites at which events took place. Relevant sites include, for example: military installations and towns, prisons and prison camps, partisan bunkers, execution sites, secret police offices, sculptures and architecture or simple squares or buildings where memorable events took place. (Baltic Initiative and Network n.d.)

In this light, the Initiative is devoted to the preservation and conservation of various natural and built landscapes and sites that hold historical significance for cultivating a formally networked Cold War psychogeography across the Baltic region.

The Initiative also devotes considerable attention to its virtual presence courtesy of a resource-intensive website (supported by the Danish Ministry of Children and Education and funded through football pools and the national lottery). It functions as a central repository of information—relating to permanent museums, temporary exhibitions, and eyewitness testimonies—from across the region. In 2010, the Initiative joined forces with the Archipelago Museum (Øhavsmuseet) on Langelands, Denmark, to sponsor a major publication entitled *Travel Guide: Traces of the Cold War* (Rasmussen 2010). Compiled by Rasmussen, the image-rich resource documents 125 relics, sites, and museums across the eleven countries. Its scope reaches beyond museums and monuments in the former communist east to encompass other structures that carry traces of the Cold War era, such as buildings that used to house KGB headquarters or host party officials on their annual holidays; western sites predominantly are former military installations and museums.

While there may be a concerted effort by the Initiative to give even and objective treatment to Cold War memories across the entire Baltic region, there are instances where it provides a welcome moral high-ground—one that might be less clear if it included memories from the Second World War. To cite one example, in writing the foreword to the aforementioned 2010 *Travel Guide*, Denmark's then-Minister for the Interior and Health, Bertel Haarder, encouraged the Initiative's wide-reaching coordination of historical consciousness across northern Europe, but also made it clear that such efforts should be underpinned by raising awareness of the legacies of communist regimes:

> This guide book has a special objective that I fully support: to use the history of the Cold War period to create historical awareness and greater mutual understanding between the countries around the Baltic Sea. This understanding is a prerequisite to more fruitful international cooperation between people, organizations, companies and international associations.
>
> It is now 20 years since the Berlin Wall fell and the Soviet Union collapsed. This leads me to raise the issue of whether people, and especially the youth of today, for example in the Scandinavian countries, have fully understood what this period meant to world history. Formerly independent nations such as Estonia, Latvia, Lithuania and Poland (all only 1 or 2 hours from Scandinavia by air) spent nearly half a century under Communist dictatorships.

> Tens of thousands of families from these countries were deported to remote parts of the Soviet Union with the aim of destroying any potential resistance in those countries. Others were executed or sentenced to long periods in prisons or concentration camps simply because their political views were different from those of the rulers. We must not forget what happened to our neighbours, just a few decades ago. (Haarder 2010, 3)

Haarder clearly expressed concern that, two decades on, memories of the Cold War era were at risk of receding into the background across northern Europe, perhaps leading to future generations not understanding the crimes of communism. Conservative anxiety thereby mixed with heritage tourism to promote shared and common memories of the Cold War throughout the Baltic region.

In more recent developments, in 2018 the Initiative took further steps toward cementing its formal status as an organization with a constituted Board and Articles of Association. It has participated in various projects with museums and universities from across its member states (Rasmussen, personal correspondence with the authors, 2019). The Initiative also continues its work in not only protecting Cold War era sites but also opening them to public access, although such efforts often prove challenging. It has facilitated tourist trips in recent years, particularly with Danish teachers visiting Cold War sites in the Baltic states. After some of its earlier conferences arranged with cultural advisors, heritage experts, tourist authorities, and academics had been staged in places such as Gdańsk, Berlin, Riga, and Tallinn, the Initiative's major conference held at Hamburg's International Maritime Museum was attended by virtually all maritime museums across the Baltic region. And, finally, in a sign that the Initiative is looking to expand its horizons, at the time of writing there are plans afoot to involve other European countries in some of its activities. Drawing on the earlier experiences of Baltic museums in collecting artefacts during and after the collapse of the Soviet Union, the Initiative is encouraging Ukrainian museums to likewise collect items from the ongoing demonstrations and struggles caused by Russia's invasion of part of the country. In widening its scope in such a way, the Baltic Initiative and Network reminds us that memory work is a fluid process, and subsequently, remembering the Cold War era at the regional level invites reimagining in dynamic ways.

Nuclear Colonialism

In the Asia-Pacific region, we see the rise of memories of the Cold War that are common by virtue of cognate experiences forming the basis of interconnected and fluid remembering. At the same time, such memories are also accretive, building on narratives of suffering, displacement and dislocation that often began in earlier forms of war and/or colonialism. For example, in the Pacific, the Marshall Islanders, while remembering the legacies of US atomic testing in 1946–58, and since the 1950s of successive forms of missile testing, recall physical damage to their environment and their displacement and service as laborers; but they do so as part of a continuum that begins much earlier with Japanese colonization and their occupation by first the Japanese and then the Americans during the course of the Second World War (Dvorak 2018). As is outlined further below, a history of colonization that highlights recent injustices is essential to the status of victims and to establishing the grounds for compensation for past suffering. Framing the recent experiences as "nuclear colonization" also creates a means for linking to the raw memories of other countries where people have endured nuclear detonations.

As a guide to conceptualizing this phenomenon, a classic of Cold War literature is a good starting point. In the mid-1970s, Aleksandr Solzhenitsyn published his famous trilogy about the tens of millions of Soviet citizens sent to forced labor camps, *The Gulag Archipelago 1918–1956: An Experiment in Literary Investigation* (Solzhenitsyn 1974, 1975, 1978). These camps became known for the acronym of the administrative organ that ran them, GULag (from *Glavnoe Upravlenie Lagerei*, "Main Camp Administration"). The Gulag represented for Solzhenitsyn thousands of islands scattered across the Soviet Union but connected strongly as a "country of Gulag" in a psychological sense; "an almost invisible, almost imperceptible country inhabited by the zek people [prisoners]" (1974, ix–x). Since the mid-1970s, *Archipelago* has retained its interpretive value for those interested in the legacies of the Cold War, partly for its capacity to capture geographical dispersal in a connected way, including sometimes in the literal geographical sense of chains of islands. North Russian islands that once hosted forced labor camps continue to draw researchers for their isolated-but-connected qualities, their layers of dark history and their value as microcosms of bigger stories (Voronina and Bogumił 2018). But *Archipelago* is also effective more broadly in its suggestive psychological, metaphorical reach.

The so-called "spatial turn" in humanities and social science research over the last two decades has enhanced the interpretive attraction of archipelagos. As a method of avoiding power-laden narratives of colonization, modernity or development, and the binary positions that flow from these, the spatial turn encourages thinking in relational ways, emphasizing mobilities, fluidity, assemblages. This way of viewing works well with islands, allowing them greater significance than the more geopolitical rendering of them as appendices to, and between, continents by taking interest in the experiential interconnections between them. To use the Marshall Islands example again, memories of US atomic testing there at Bikini and Enewetak connect to the remembering of other island episodes of testing such as the French at Mururoa Atoll and the Anglo-Australian ventures on the big island of Australia. Scholars writing in the relatively new *Journal of Island Studies*, following Deleuze and Guattari (1986) and De Landa (2006) among others, suggest that archipelagos can frame assemblages wherein diverse component pieces connect in ways that suggest distinctive forms at the level of a whole (Pugh 2013). The physical, geographical meaning of island is accompanied by an interpretive act of constant joining up with other islands in order to reveal bigger stories with common roots. To what extent does this help our understanding of remembering the Cold War in island settings? While we are not the first to explore the value of an assemblage approach to aspects of the Cold War (see Shringarpure 2019) we suggest here that the archipelago-assemblage lens offers a valuable entry point to some of the Cold War's mnemonic consequences in the Asia-Pacific region. Indeed, we find that this process is underway: there are increasing signs of others building such a sense of mnemonic connectedness. The Cold War in the Pacific is remembered, for reasons going to common experiences and victim campaigns for compensation under the heading of "nuclear colonialism"; and in other parts of the region we find a less victim-focused and more economically-positioned remembering responding to a history of militarization and a growing industry in heritage tourism.

Where Solzhenitsyn's archipelago inhabitants were joined by the common experience and remembering of hunger, death and privations in labor camps, one common feature of Cold War suffering is that of the consequences of nuclear colonialism—the Cold War testing of weapons in less populated land regions and islands where the interests of local peoples were largely ignored. Memories of these experiences, in different parts of the world, have become more thoroughly documented and more joined up in recent times. In part, the added momentum to remembering nuclear colonialism stems from recalling anniversaries (including the seventieth, sixtieth and fiftieth anniversaries of the start of testing at different sites), from renewed compensation claims for the health and economic consequences of testing as well as forced relocation, and from the aging of affected populations. Second generations and historians have been eager to collect memories of these traumatic times while they still can. This includes not only local populations but also the servicemen of nuclear powers such as Britain, France and the United States, who helped conduct the tests.

In telling these stories, both government archives of the colonial states and the personal testimonies of those who were present at or near the testing are important. The colonial powers' records have been valuable when documenting the detailed nature of events, but access to them varies as governments have enforced secrecy and restrictions on information, so they cannot be relied on to provide all the forms of evidence that might be needed to make claims of compensation. Thus, oral histories and memoirs are among the forms of intensified remembering taking shape, while researchers prise open state secrets, such as they can, to add more knowledge to case studies (Trundle 2011).

The basis for this activity is the fifty years of post-war testing nuclear weapons in outposts of empire between 1946 and 1996, beginning with the United States in the Marshall Islands and concluding with the French test at Mururoa Atoll (and the Chinese test at Lop Nur in the Xinjing Uyger Autonomous Province). Including the tests in 1945 leading up to the bombing of Hiroshima and Nagasaki, more than 2000 tests were carried out across the world up to 1996, when a Comprehensive Nuclear Test Ban Treaty curbed the activities of many, but not all, of the nuclear-capable powers. What constitutes a "test" remains open to debate, as some of the smaller experimental explosions were "dirtier" than major tests. According to the Comprehensive Test Ban Treaty Organization, using a high explosive measure, the Pacific was particularly affected, with the Americans conducting 111 tests and the French conducting 193 tests in and near Pacific islands during the fifty-year period (Comprehensive Nuclear-Test-Ban-Treaty Organization 2019).

The sense of grievance for many islanders remains raw. Bikini Islanders from the Marshall Islands, for example, remain excluded from their home, where US hydrogen bomb testing occurred in the 1950s. Some locals returned briefly when the atoll was declared safe in the 1970s, before being relocated again in 1978 when tests revealed dangerous levels of radiation. The long-term impact of US testing in the Marshall Islands continues to cruel relations with now-independent Marshall Islanders. In common with French Polynesian Islands where testing occurred, the aftermath in the Marshalls is seen in illnesses, premature deaths and birth defects, according to local populations, but also includes the shrinkage of living space as parts of archipelagos are militarized, creating small urban ghettoes of unsanitary living and greater dependence on expensive food products needing to be imported (Dvorak 2018; Firth 1997) (Fig. 3).

In the case of the Marshall Islands, in 1983, Washington agreed to provide Islanders with US$150 million for damages caused by the tests conducted at Bikini and Enewetak atolls. In 2000, the Marshall Islanders sought an increase of the moneys owed through changed circumstances. By the mid-2000s, investigations by a claims tribunal had revealed the need for around another $2 billion, including housing and social needs, but the fund was exhausted (Okney 2010) and US authorities have resisted calls for more. While some locals have benefited financially from the US militarization of their islands, for most Marshall Islanders the balance sheet remains grim; and they now face the twin perils of rising sea levels through climate change and a decaying and water-affected concrete dome

Fig. 3 US nuclear weapons test at Bikini Atoll, Marshall Islands, 1946. (Source: The International Campaign to Abolish Nuclear Weapons, https://www.flickr.com/photos/30835738@N03/7936491790. Credit: US Government. Licensed under CC BY-NC 2.0)

over a bunker loaded with radioactive materials from the testing. The other major US military activity in the islands, missile testing, continues today. The Ronald Reagan Ballistic Missile Defense Test Site provides for ballistic missile testing and space operations on Kwajalein Atoll and Wake Island.

While drawing on stories of particular times and places, the remembering of atomic testing in French Polynesia (forty-four atmospheric tests at the South Pacific atolls of Mururoa and Fangataufa between 1966 and 1974, then 149 underground tests to 1996) and in the Australian desert and offshore islands (twelve major tests and many more "minor" ones between 1952 and 1963) bear similar features for their recollection of the suffering, dislocation and continued injustice. To some extent, recalling this activity and its activities is defining the Pacific. As Matt Matsuda has shown, the idea of the Pacific itself is very much shaped by the questions being asked about this region of the world and the disciplinary lenses brought to bear (Matsuda 2012). Whether a site of nuclear colonialism or of "transpacific nuclear modernity" (Cho 2019), the Pacific has grown as a canvas on which differently located memories can be gathered and joined. In addition to the French and US tests mentioned, the British carried out nine major tests, in the late 1950s, in islands of what were the British Gilbert and Ellice Islands Colony, Malden and Christmas Island, which today form part of independent Kiribati. The Americans followed later by conducting twenty-four tests at Christmas Island. One study observing the efforts of those servicemen involved in the British tests of hydrogen bombs at Christmas and Malden Islands noted the circulation and overlapping of stories and evidence across national boundaries, as the different groups of servicemen involved—22,000 British, 14,000 Australian, 500 New Zealand and 150 Fijian—gathered different forms of evidence to use in ongoing claims for compensation for health problems subsequently endured (Trundle 2011). Like most servicemen who participated in nuclear testing, they have been denied compensation, with British governments insisting that they did not suffer from unusual exposure to radioactive materials and that any subsequent illnesses could not be linked to their earlier involvement in the tests. Where compensation has been provided for the consequences of nuclear testing in the Pacific, it has generally been regarded as too low in sum and too restricted in scope.

The joining up of different parts of the atomic testing archipelago happens also through the work of museums and educational institutions. The Hiroshima Peace Memorial Museum connects a display about the treatment of scarred survivors from the Hiroshima bomb, with the story of the Japanese fishing boat ***Fukuryu-maru No. 5*** that was covered with radioactive fallout near Bikini Atoll in March 1954, as a result of a huge US nuclear test of a Hydrogen or H-bomb. "One of the crew members died later that autumn," reads the caption, "from symptoms similar to those common in Hiroshima and Nagasaki" (Hiroshima Peace Memorial Museum 2013). And, not surprisingly, given the size of the H-bomb detonation and the number of fishing vessels passing through that mid-Pacific region in the aftermath, the lens of concerned inquiry has broadened beyond Hiroshima. School collaborations between Japan and South Korea since the 1990s have led to greater knowledge about Korean fisherman being affected by radiation, and of a connection to Taiwan, where the Koreans purchased recycled Japanese vessels better suited to their needs.

Educational exchange, pen-pal connections and the "twinning" of nuclear-affected cities has occurred between the United States and former Soviet sites in Kazakhstan and Russia (Lowe and Joel 2013). Since the mid-1980s the emergence of a South Pacific anti-nuclear movement has encouraged greater connectedness between affected islands at political as well as cultural/educational levels. Initially aimed mostly at the French who continued to test nuclear bombs, the focus of the movement has broadened to encompass all episodes of nuclear testing and the many unresolved consequences (Firth 1997; Johnson 2018). In recent years the Nuclear Free and Independent Pacific Movement has united victim islanders and their descendants whose common causes include: acknowledgment by governments of their responsibility for impacts of nuclear testing; greater openness and assistance for those seeking details about the tests and impacts; and improved levels of compensation, completion of further clean-ups needed, and recurrent monitoring of affected sites (Maclellan 2005).

Military History/Heritage Tourism

The island of Okinawa is one of 160 islands making up the Ryukyu Archipelago making up the Prefecture of Okinawa in the southernmost part of Japan. It is a little over 1500 kilometers north-east from Naha, the Okinawan capital, to Tokyo. In the opposite direction, south-west, it is only a little over 600 kilometers to the Taiwanese capital of Taipei and continuing south west only a little under 1000 kilometers from Naha to the Taiwanese island of Kinmen, dubbed the "Cold War Island" by historian Michael Szonyi, for its status as the frontline of China-Taiwan military and propaganda sorties (Szonyi 2008). At less than three kilometers from the coast of mainland China, and comfortably within artillery range, Kinmen was literally the frontline of the Cold War in East Asia. For its part, Okinawa, while often remembered as the site of particularly bloody battles in the last stages of the Second World War, has a strong Cold War identity. After the war and after the ending of the American occupation of Japan in 1952, Okinawa remained occupied until its so-called "reversion" to Japan in 1972; and thereafter it has continued to host major US airbases.

Okinawa and Kinmen constitute examples of archipelago regional remembering in three connected ways. The first is by virtue of their historical and geo-political circumstances in past and contemporary state contests in East Asia. They sit in a region in which contemporary power struggles are based on claims to sovereignty complicated by, and requiring reference to, the Cold War. Indeed, this struggle connected them materially up to the late 1980s. During the Cold War, much of the US military assistance provided to Taiwan, including Kinmen, came from Okinawa. In the longer historical view, Okinawa is the largest of the chain of 55 islands (and many smaller islets) stretching from near the coast of the southern Japanese island of Kyushu to near the northern edge of Taiwan. This chain formerly constituted the Ryukyu kingdom, a Chinese tributary state for most of the fourteenth and fifteenth centuries, but then tied loosely to both China and Japan until 1879, when Japan annexed the kingdom. One uninhabited group of islands within the chain, called the Diaoyu islands by China (or Diaoyutai by Taiwan) and the Senkaku islands by Japan, has been the subject of unresolved contestation between the People's Republic of China, Japan and the Republic of China (Taiwan) since the 1970s.

It was the 1971 Okinawa Reversion Agreement handing back Okinawa to Japan the following year that emboldened a Japanese claim for sovereignty over the islands and triggered the new wave of disputation. The Diaoyu/Senkaku Islands sit roughly half-way between Okinawa and Taiwan; and their disputed status is among the most intractable of problems bedeviling Sino-Japanese relations today. Acts of remembering that invoke victimhood and injustice are fundamental components of claims to sovereign rights by the three parties. For example, in 2018 the Japanese opened a Museum of Territory and Sovereignty to set out their historic claims to Senkaku and another island disputed by South Korea (National Museum of Territory and Sovereignty 2020). In the same year, China opened its own South China Sea Museum on Hainan Island, full of materials pointing to its long-established rights over Diaoyu, and rejecting Japanese claims linked to the history of US occupation of Okinawa (*Beijing Review* 2016). In both cases, artefacts and other sources of heritage, and accounts of Cold War events, convey state-constructed history designed to influence international law and opinion, and to foster more popular appreciation of injustices wrought by wartime circumstances.

Secondly, both Okinawa and Kinmen are small and distant parts of large island nation-states, the governments of which have been ready to sacrifice local, Okinawan and Kinmen interests for the sake of bigger strategic-economic priorities. As a result, local leaders on both islands have been proactive in the post-Cold War era in lobbying and agitating for change. Negotiating local interests in a sea of ongoing contests for power is the common theme. In the case of Okinawan politicians, this has included bursts of activism attempting to mark out a distinctive cultural identity alongside Japanese citizenship, promoting connections with Indigenous minorities in other countries, and adventurous thinking around economic futures that would enable greater independence from Tokyo (Hein and Selden 2003). In Kinmen, local politicians campaigned successfully to remove the garrison that had propped up Kinmen's economy up to the 1980s, and then sought to fill the gap (Szonyi 2008).

And the third and closely related point is that they both achieved exaggerated strategic significance on account of the Cold War, to the extent that they both depended on military activities as the basis of their respective economies. To some extent, this situation still pertains for Okinawa, as is discussed further below, but leaders on both islands have, in recent years, tried to think hard about how a future economy based on tourism, including Cold War tourism, might be sustained. Both Okinawa and Kinmen have turned to tourism in ways that are partly entwined with, and seek partly to keep separable, their military pasts. In needing to manage this tension, and needing to balance their own plans for tourism with the directives of Tokyo and Taipei, respectively, they are to the fore of a nimble form of heritage tourism that is unfolding in different parts of the region. A heritage tourism that is enabling a loose sense of regional Cold War tourism to take root. The short discussion below expands on these three related themes.

Since 1994 when they opened for travelers, the Taiwanese islands of Kinmen (formerly Quemoy) and the Matsu Archipelago, located in the northwest Taiwan Strait—Matsu is only slightly further away from the

China mainland at eight kilometers—have been increasingly promoted as logical tourist destinations. Until 1994 the islands were under a war zone administration. They had become part of Chiang Kai-shek's Republic of China (ROC) when he with 600,000 of his remaining forces set up government in Taiwan in 1949. Later that year, Chiang's troops fought off an attempt by Mao to seize Kinmen. When the Korean War broke out on June 25, 1950, US support for the independent Taiwan firmed and President Truman declared the Straits of Formosa to be neutral, to be patrolled by the US Seventh Fleet. Chiang militarized both Kinmen and Matsu both defensively and in ways that provided for a possible offensive against the mainland. He also sent successive waves of national servicemen to bolster their defenses. The islands were an international flashpoint during 1954–55 and 1958–59, when, in spikes of PRC-ROC jousting, artillery fire from the mainland screamed down on the inhabitants. In 1960, policy towards the islands featured in the US Presidential debates between candidates John F. Kennedy and Richard Nixon. The couplet "Quemoy and Matsu" became a Cold War refrain and harbinger of how a third world war might start. From that time, however, relations between ROC troops and the mainland settled into the unusual ritual of mainland artillery shelling every second day (of shells filled with propaganda rather than high explosive), in return for the US Seventh Fleet keeping its distance. All shelling stopped in 1979.

Both Matsu and Kinmen have made a rapid transition in last twenty years from militarized identity to tourist destination, with Taipei's encouragement. Focusing here on Kinmen, this includes large groups of tourists arriving both from Taiwan's main island (noting that Taipei is some 300 kilometers away) and from nearby Xiamen City in Fujian Province on the Chinese mainland. Kinmen opened to mainland visits in 2005. Although fluctuations in the state of Taipei-Beijing relations have the potential to turn the mainland tourism tap off, recent annual numbers of visitors (both mainland and Taiwanese) have been close to half a million (Chen and Tsai 2019).

What do they come to see and do? The military tunnels and coastal defenses of the main island of Greater Kinmen have been a drawcard. The tunnels feature a waterway that was to shelter landing craft prior to a possible invasion of the mainland. Elsewhere a military museum features a jeep in which Chiang Kai-shek toured the island after the invading communists had been seen off in 1949. Entrepreneurial craftsmen have found a niche in selling cleavers and knives fashioned out of old artillery shells.

Kinmen's militarized past has been successfully commodified, even playfully so, taking into account such delicacies as "Battlefield Mine Cakes" that will "explode your taste buds" (Zhang 2017). The tourist industry on Kinmen has been dynamic, experimenting with a mix of experiences and tours. Notably, the military past is also folded into a bigger tourist narrative that features distinctive and traditional Chinese architecture, wetlands and other natural beauty, spectacular beaches and local crafts and artistic products (Kinmen Travel 2019). A Kinmen National Park, established in 1995, embraces war and natural heritage, and the official tourist website boasts that Kinmen is "Taiwan's top potential UNESCO World Heritage Status site" (Kinmen Travel 2019). Indeed, Kinmen is seeking World Heritage listing, with some scholarly interest and support. The Cold War provides much of the glue that binds its attractions together. As Szonyi writes: "If authenticity means faithfulness to an inherited set of practices reproduced in a specific locale for specific purposes unconnected to the market process, authenticity in Jinmen [or Kinmen] is distinguished by its link not to the pre-market past but to the geopolitical past" (Szonyi 2008, 210) (Figs. 4 and 5).

The transition to tourism has not eased the memories of suffering under the military past for some. Kinmen's residents have not completely dropped their tales of hardships, enabling some to gain compensation from accidents and appropriations by the ROC government (Szonyi 2008, 236–37). But these aspects of Kinmen's past have also been tempered by the UNESCO-friendly tourist vista. A recent study of tourist visitors from both directions, the Chinese mainland and Taiwan's main island, suggest a high degree of commonality in motives for wanting to visit Kinmen: both Chinese mainlanders and Taiwanese indicated similar spiritual and experiential approaches to walking the battlefield sites and encountering other forms of cultural tourism (Chen and Tsai 2019).

The embrace of tourism is also the way forward for Japan's Cold War Island of Okinawa. In 1995 the Prefecture Governor of Okinawa, Ota Masahide, pronounced that Okinawa would be Japan's tourism prefecture. A vocal opponent of the US bases that have been so fundamental to the prefecture's economy, Masahide looked forward to the day when tourism would grow to such proportions that it would replace the Americans in its economic benefits (Figal 2012, 113–17). Masahide was not alone in his longing for the day that Okinawa could proudly stand without this enduring form of foreign occupation. The Okinawan Prefectural Museum celebrates its heritage as the center of the Ryukyu kingdom before being

Fig. 4 Kinmen, Zhai Shan tunnel. (Source: uniquenikki, https://www.flickr.com/photos/15184586@N08, licensed under CC BY-ND 2.0)

forcibly absorbed into Meiji Japan in 1879. Since that time, according to the museum displays, Okinawan inhabitants were dragged into Japan's militarism, which led to the island experiencing the most horrific of the Pacific battles of the Second World War and (mainland) Japanese military commanders then forcing the locals into acts of mass suicide after the Americans invaded in 1945. Okinawans had to endure American administration until reversion in 1972 (Okinawa Prefectural Museum and Art Museum 2013). During that time, the island's strategic importance to the United States in the Pacific was underlined by the Korean War and then American intervention in Vietnam. The US bases on Okinawa played crucial roles in both conflicts. Since 1972, with Tokyo's agreement, the Americans have maintained nearly 30,000 military personnel on the islands at several bases, constituting 75 per cent of all US forces in Japan.

Okinawa's economy has become increasingly dependent on the Americans; but calls for removal of the bases have grown and in February 2019 a big majority of residents voted against relocation of the big airbase of Futenma. Rather than relocate it from its current site in a highly

Fig. 5 Kinmen, Master Wu forging knives from old artillery shells. (Source: ipickmynose, https://www.flickr.com/photos/27275195@N05/4247039816, licensed under CC BY-NC-ND 2.0)

developed residential area, residents wanted it gone; but their vote is not binding on Tokyo, and the future of the base will remain a source of tension between Okinawans and Tokyo/Washington. Acts of violence have played a role in stirring local opposition, including three US servicemen's rape of a twelve-year-old girl in 1995.

The turn to tourism on Okinawa had begun before Masahide's call in 1995, and it has unfolded in a manner that has been partly dependent on the island's military past and present. As Gerard Figal's detailed study shows, even during US occupation, Okinawans were adapting their landscape to meet tourist preferences, including the planting of flowers on sites of mass killings from the Second World War, and introducing new forms of vegetation similar to those found in Hawaii, and more attuned to mainland Japanese expectations of the Pacific island idyll (Figal 2012). Since reversion in 1972, Okinawa has showcased its rich cultural past of the Ryukyu kingdom, including its cultural ties to China, as well as expanding on the natural and built attributes of paradise in the Pacific. At the same time, however, the nine million annual tourists (Japan Times 2018), overwhelmingly from the main Japanese islands, have continued to regard the US bases, observing aircraft take-offs and wandering through the American shopping village, as part of the tourist experience. The strongest messaging from Okinawan tourism might have encouraged looking beyond the island's Second World War and Cold War identity, but visitors continue to include it (Figal 2012). Envisaging a future in which tourism on Okinawa replaces the current economic role of US airbases is therefore all the more complicated.

One tourism scholar, William Cannon Hunter, examined both Okinawa and Kinmen comparatively, for their transitions from war to tourism economies, and for how they managed their common dependence on nearby larger populations. He found much in common between the two sites as they embraced tourism as their future (Hunter 2012). The short analysis here suggests that remembering the Cold War's local and regional impact matters to Kinmen and Okinawa in ways that extend also to ongoing power struggles in the South China Sea, and ongoing efforts to chart distinctive futures as small islands within bigger nation-states that have proven willing to override their interests.

Whether or not Kinmen succeeds in its UNESCO heritage listing, Kinmen, Okinawa, and other parts of a Cold War inflected archipelago are engaging in the complex politics and diplomacy of cultural heritage with determination and some deftness. Their acts of remembering are largely common in form—different groups are joining to remember forms of Cold War militarism as cognate experiences in ways that preserve their special claims as victims or agitators for change while drawing on the broader narrative of regional Cold War. The sense of region is necessarily flexible, but the global Cold War provides easily for this, with histories of occupation and particular spikes in hostilities foregrounding memories of layered oppression and suffering.

The regions of Cold War remembering examined in this chapter feature both state-centric memory work and more grass-roots popular efforts, both of which connect to political and economic opportunities. There are clear differences between the case studies. The V4 in Europe have mobilized regional remembering around new symbols and in relation to the opportunities for membership of powerful collectives in NATO and the EU. Other efforts to remember regionally are more iterative and join up popular interest with state initiatives as opportunities arise, as in the cases of the Baltic Initiative and Network and the coalition of activism remembering nuclear colonialism in the Pacific. But all cases presented here are sensitive to changing politics, and the fickle dynamics of support from government agencies, community organizations and local champions. In the cases of the Baltic Initiative and Kinmen and Okinawa, the meshing of Cold War with the tourist-heritage economies growing in Europe and in Asia and the Pacific is an evolving story (see Winter 2014). The assemblages and shifting emphases of the past testify to the complex interplay of victims versus conciliation, and political cause versus commercial opportunity. In all cases, the acts of remembering in local, regional and global frames continue to evolve according to identified needs of the near future, even as they appeal to shared and common memories of the Cold War past.

References

Ashplant, T. G., Graham Dawson and Michael Roper. 2000. The Politics of War Memory and Commemoration: Contexts, Structures and Dynamics. In *The Politics of War Memory and Commemoration*, ed. T. G. Ashplant, Graham Dawson and Michael Roper, 3–86. London/New York: Routledge.

Baltic Initiative and Network. n.d. The Idea and Objectives. http://coldwarsites.net/network/. Accessed October 4, 2019.

Bárd, Petra, and Anna Śledzińska-Simon. 2019. Rule of Law Infringement Procedures: A Proposal to Extend the EU's Rule of Law Toolbox. *CEPS Paper in Liberty and Security in Europe*, no. 2019–09. https://www.ceps.eu/wp-content/uploads/2019/05/LSE-2019-09_ENGAGE-II-Rule-of-Law-infringement-procedures.pdf.

BBC News. 2018. Karl Marx Statue from China adds to German Angst. May 5, 2018. https://www.bbc.com/news/world-europe-44009621.

Beijing Review. 2016. "South China Sea Museum Set to Open in Hainan". November 24. http://www.bjreview.com/Latest_Headlines/201611/t20161124_800072545.html

Birnbaum, Ben. 2011. Statue in Budapest's Liberty Square credits Reagan for Freedom. *The Washington Times*. June 29, 2011. https://www.washingtontimes.com/news/2011/jun/29/statue-in-budapests-liberty-square-credits-reagan-/.

Chen, Chen-Min, and Tsung-Hsien Tsai. 2019. Tourist motivations in relation to a battlefield: a case study of Kinmen. *Tourism Geographies* 21, no. 1: 78–101.

Cho, Yu-Fang. 2019. Remembering Lucky Dragon, Re-membering Bikini: Worlding the Anthropocene through Transpacific Nuclear Modernity. *Cultural Studies* 33, no. 1: 122–46.

Comprehensive Nuclear-Test-Ban-Treaty Organization. 2019. https://www.ctbto.org/nuclear-testing/history-of-nuclear-testing/world-overview/.

De Landa, Manuel. 2006. *A New Philosophy of Society: Assemblage Theory and Social Complexity*. London: Continuum.

Deleuze, Gilles, and Felix Guattari. 1986. *Kafka: Towards a Minor Literature*. Minneapolis: University of Minnesota Press.

Dvorak, Greg. 2018. *Coral and Concrete: Remembering Kwajalein Atoll between Japan, America, and the Marshall Islands*. Honolulu: University of Hawaii Press.

Embassy of the United States, Prague. 2011. Czech Republic Pays Tribute to Ronald Reagan. July 1, 2001. http://prague.usembassy.gov/ronald-reagan-street.html.

Figal, Gerard. 2012. *Beachheads: War, Peace and Tourism in Postwar Okinawa*. Lanham, MA: Rowman and Littlefield.

Firth, Stewart. 1997. A Nuclear Pacific. In *The Cambridge History of the Pacific Islanders*, ed. Donald Denoon with Stewart Firth, Jocelyn Linnekin, Malam Maleiia and Karen Nero, 325–58. Cambridge: Cambridge University Press.

Gera, Vanessa. 2011. Walesa unveils Statue of Ronald Reagan in Warsaw. *Boston.com*, November 21. http://archive.boston.com/news/world/europe/articles/2011/11/21/walesa_unveils_statue_of_ronald_reagan_in_warsaw/.

Gyárfášová, Ol'ga, and Grigorij Mesežnikov. 2016. *25 Years of the V4 as Seen by the Public*. Bratislava: Institute for Public Affairs.

Haarder, Bertel. 2010. Foreword. In *Travel Guide: Traces of the Cold War Period. The Countries around the Baltic Sea*, ed. Johannes Bach Rasmussen, 3. Copenhagen: Nordic Council of Ministers.

Havel, Václav. 2007. *To the Castle and Back*, trans. Paul Wilson. New York: Alfred A. Knopf.

Hawley, Charles. 2009. Spiegel Online Interview with Lech Walesa: "It's Good that Gorbachev was a Weak Politician". *Spiegel Online International*. November 6. https://www.spiegel.de/international/europe/spiegel-online-interview-with-lech-walesa-it-s-good-that-gorbachev-was-a-weak-politician-a-659752.html.

Hein, Laura and Mark Selden, eds. 2003. *Islands of Discontent: Okinawan Responses to Japanese and American Power*. Lanham, MA: Rowman and Littlefield.

Hiroshima Peace Memorial Museum. 2013. Exhibit Text.

Hungary Today. 2019. Orbán: "Juncker shouldn't be Educating Us on Xenophobia". *Hungary Today*. April 2. https://hungarytoday.hu/orban-juncker-shouldnt-be-educating-us-on-xenophobia/.

Hunter, William Cannon. 2012. The Good Souvenir: Representations of Okinawa and Kinmen Islands in Asia. *Journal of Sustainable Tourism* 20, no. 1: 81–99.

Japan Times. 2018. Okinawa tourist numbers top those of Hawaii for first time. 9 February. https://www.japantimes.co.jp/news/2018/02/09/national/okinawa-tourist-numbers-top-hawaii-first-time/

Johnson, Giff. 2018. Marshall Islands assured US Compact and nuke ban compatible. Radio New Zealand. https://www.rnz.co.nz/international/pacific-news/361219/marshall-islands-assured-us-compact-and-nuke-ban-compatible.

Johnstone, Chris. 2011. Ronald Reagan to be Honored in Czech Capital. *Česká pozice*. http://ceskapozice.lidovky.cz/tema/ronald-reagan-to-be-honored-in-czech-capital.A110630_143237_pozice_23468. Accessed June 30, 2011.

Judt, Tony. 2010. *Postwar: A History of Europe Since 1945*. London: Vintage.

Kinmen Travel. 2019. https://kinmen.travel/en/discover/kinmen. Accessed August 6, 2019.

Klein, Rahel. 2019. Europe: This is How the Visegrád Group Works. *Deutsche Welle*. https://www.dw.com/en/this-is-how-the-visegrad-group-works/a-47402724. Accessed February 7, 2019.

Lowe, David, and Tony Joel. 2013. *Remembering the Cold War: Global Conflict and National Stories*. Abingdon: Routledge.

Maclellan, Nic. 2005. The Nuclear Age in the Pacific Islands. *The Contemporary Pacific* 17, no. 2: 363–72.

Matsuda, Matt. K. 2012. *Pacific Worlds: A History of Seas, Peoples and Cultures*. Cambridge: Cambridge University Press.

National Museum of Territory and Sovereignty (Chiyoda-ku, Tokyo, Japan). 2020. https://www.cas.go.jp/jp/ryodo_eg/tenjikan/senkaku.html. Exhibit text.

Okinawa Prefectural Museum and Art Museum. 2013. Exhibit text.

Okney, Philip A. 2010. Legacies and Perils from the Perspective of the Republic of the Marshall Islands Nuclear Claims Tribunal. In *The Oceans in the Nuclear Age: Legacies and Risks* ed. David D. Caron and Harry N. Schieber, 49–67. Leiden: Martinus Nijhoff.

Pugh, Jonathan. 2013. Island Movements: Thinking with the Archipelago. *Island Studies Journal* 8, no. 1: 9–24.

Rasmussen, Johannes Bach. 2010. *Travel Guide: Traces of the Cold War Period. The Countries around the Baltic Sea*. Copenhagen: Nordic Council of Ministers.

Shringarpure, B. 2019. *Cold War Assemblages: Decolonization to Digital*. New York: Routledge.

Solzhenitsyn, Aleksandr. 1974, 1975, 1978. *The Gulag Archipelago 1918–1956: An Experiment in Literary Investigation (3 Vols)*. Trans. Thomas P. Whitney. London: Collins.

Stone, Jon. 2018. "Today He stands for Things, which He is Not Responsible for": EU President Juncker Defends Karl Marx's Legacy. *Independent*, May 4.

https://www.independent.co.uk/news/uk/politics/karl-marx-jean-claude-juncker-defends-legacy-a8337176.html.

Szonyi, Michael. 2008. *Cold War Island: Quemoy on the Front Line*. Cambridge: Cambridge University Press.

Terenzani, Michaela. 2010. A Common but Contentious History. *The Slovak Spectator*. December 13. https://spectator.sme.sk/c/20038369/a-common-but-contentious-history.html.

Trundle, Catherine. 2011. Searching of Culpability in the Archives: Commonwealth Nuclear Test Veterans' Claims for Compensation. *History and Anthropology* 22, no. 4: 497–512.

Visegrád Group. 2018. Fidesz MEPs Call on Juncker to Cancel Speech on Marx Anniversary. May 3. http://www.visegradgroup.eu/news/fidesz-meps-call-on.

———. n.d. History of the Visegrád Group. http://www.visegradgroup.eu/about/history. Accessed September 28, 2019.

Vondra, Alexandr. 2018. Cold War between Visegrád and Brussels. *Geopolitical Intelligence Services*. May 31, 2018. https://www.gisreportsonline.com/cold-war-between-visegrad-and-brussels,politics,2568,report.html.

Voronina, Tatiana, and Zuzanna Bogumil. 2018. Islands of One Archipelago: Narratives About the Solovetskie Islands and the Memory of Soviet Repressions. *Laboratorium: Russian Review of Social Research* 10, no, 2: 104–21.

Winter, Jay, and Emmanuel Sivan. 1999. Setting the Framework. In *War and Remembrance in the Twentieth Century*, ed. Jay Winter and Emmanuel Sivan, 6–39. Cambridge: Cambridge University Press.

Winter, Tim. 2014. Beyond Eurocentrism? Heritage Conservation and the Politics of Difference. *International Journal of Heritage Studies* 20, no. 2: 123–37.

Zhang, J.J. 2017. Rethinking "Heritage" in Post-conflict Tourism. *Annals of Tourism Research* 66: 194–96.

Human Rights and Regions of Memory: The Case of the International People's Tribunal on Crimes Against Humanity in Indonesia 1965

Katharine McGregor

Introduction

In late 2015, Indonesian human rights activists staged an international people's tribunal to hear the case of crimes against humanity, including mass murder and imprisonment without trial, committed fifty years prior

This chapter has been written with the support of the Australian Research Council Future Fellowship (FT130100957) for the project *Confronting Historical Injustice in Indonesia: Memory and Transnational Human Rights Activism*. ORCID No 0000-0001-6955-128X.

K. McGregor (✉)
School of Historical and Philosophical Studies, University of Melbourne, Parkville, VIC, Australia
e-mail: k.mcgregor@unimelb.edu.au

S. Lewis et al. (eds.), *Regions of Memory*, Palgrave Macmillan Memory Studies, https://doi.org/10.1007/978-3-030-93705-8_3

in the 1965 Indonesian genocide against the political left. The tribunal was not legally binding. Its main purpose was to increase pressure on the Indonesian government to address this case after eighteen years of failed initiatives (Marching 2016). Activists chose The Hague in the Netherlands as the location for the tribunal because holding it in Indonesia would have been too dangerous, due to ongoing threats and intimidation against activists who had taken up this case. The International People's Tribunal on Crimes against Humanity in Indonesia 1965 (hereafter IPT 1965) is beginning to receive more attention in analyses of Indonesian efforts to achieve justice for the 1965 case (Santoso and van Klinken 2017; McGregor and Setiawan 2019; Wieringa et al. 2019), but few scholars have examined it through the lens of its multiple transnational resonances. Most studies of memory and the 1965 violence furthermore examine this case from the confines of the Indonesian nation and the dynamics of memory debates between survivors and others living in Indonesia or within local communities (Kolimon et al. 2015; Kimura 2015; Kuddus 2017; McGregor and Fealy 2010; McGregor 2018).

Throughout this chapter I reflect on a range of different regions of memory invoked in memory struggles surrounding the IPT 1965. The term region generally refers to a geographic space and indeed the Cambridge dictionary defines a region as "a particular area or part of the world." When thinking about memory, however, it is now understood that memory is never really contained within one geographical space such as a nation; instead, memory is often referential or transcultural, "unfolding across and beyond cultures" (Erll 2011, 9). Yet we do not yet have precise ways of explaining how certain memory struggles are bound to or connected to others. Throughout this chapter, then, I use the concept of "regions of memory" not in a narrow geographic sense, but instead to think about the regions of memory to which Indonesia is connected through historical processes involving human rights abuses including colonization and the Cold War, as well as through parallel and intersecting human rights related struggles that have provided inspiration for Indonesians. I suggest that each of these connections brings Indonesia into different configurations of human rights related regions of memory. Human rights activists have used a variety of different strategies to draw on these human rights-related regions of memory, and to connect the 1965 case to other cases of historical injustice or symbols of historical justice.

First of all, through the choice of The Hague as the seat of international justice, the appointment of a panel of international human rights experts and lawyers and the adoption of this form of people's tribunal, activists sought to place the 1965 case on par with other cases of past human rights abuses that have received international attention and led to international prosecution or people's tribunals (Katjasungkana and Wieringa 2019). In this way they attempted to secure a place for the 1965 case in global memory of human rights injustices, linking it to other human rights struggles. Secondly, through the targets of prosecution, they invoked a historically based region of memory bounded by means of a shared commitment across several Western countries to anti-communism, including support for the violent repression of alleged communists. They did this by accusing not only the Indonesian government, but also the governments of the Netherlands, the United States and Australia of complicity in human rights crimes in Indonesia in 1965.

What activists had perhaps not fully anticipated, however, was the other human rights-related region of memory invoked in response to the tribunal. In their efforts to rebuke the tribunal, some Indonesian officials raised cases of Dutch colonial violence against Indonesians, which have received much attention in recent years. Representatives of the Indonesian government attacked the activists behind the tribunal, accusing them of being traitors for raising this case in the Netherlands, the country of the former colonial power (Hutwan 2015; McGregor and Purdey 2016). Underlining the nationalist reaction to the tribunal was the demand that this case not be internationalized and that it should instead be settled in Indonesia.

Using the case of the IPT 1965, I discuss how Indonesia's historical connections with particular countries or cases of human rights redress, based on entwined histories and parallel struggles, played out during the tribunal and the multiple human rights regions of memory at work here. I demonstrate the drive from activists to internationalize this case and make it one of supranational concern and relevance alongside the ongoing efforts to keep this case within the domestic realm. In doing so I consider the instrumentalization of regions of memory by transnational human rights activists and their detractors.

Human Rights Regions of Memory

Recent theoretical approaches to memory emphasize that the presumption that the nation is a container for memory is artificial, arguing instead that memory processes extend beyond the nation and as such should be studied in that way. To capture these more global processes, Phillips and

Reyes refer to a "global memoryscape" that is "a complex landscape upon which memories and memory practices move, come into contact and are contested by, and contest other forms of remembrance" (Phillips and Reyes 2011, 15). Astrid Erll alternatively uses the concept of "transcultural memory" to refer to a process by which memories travel and are translated to different cultural settings (Erll 2011). These are useful concepts that open up new ways of thinking about memory beyond the nation, but they do not set out what kinds of connections might be particularly prominent in memory struggles across the world. In order to think more precisely about which memory struggles intersect, the term "regions of memory" offers more specificity. It allows us to consider how one particular memory struggle is bound in different ways to other memory struggles and parts of the world, through diverse processes including historical ties, joint struggles for justice, and joint forms of oppression.

By using the novel concept of "regions of memory" and focusing on the struggle for justice for one human rights case in Indonesia, I seek to more fully capture processes of global interaction across memory struggles. Several studies have begun to point to the global flow or cross referencing of memory struggles relating to human rights cases across the world, but the most frequent reference point in these works is the Holocaust and memory originating in Europe. Rothberg's (2009) impressive work on "multidirectional memory", for example, examines how France's struggles over Holocaust memory and memory of the violent decolonization of Algeria intersect. Levy and Sznaider (2010, 6) suggest that the frequent deployment of memories of the Holocaust and related justice measures globally, especially since the end of the Cold War, represents a new form of cosmopolitanism, a way in which global concerns become more connected with local experiences or vice versa due to a global memory imperative. These are useful ways of conceptualizing memory practices, yet as I will demonstrate, there is a far broader and more complex process of exchange of memory practices and cross referencing across the globe than the spread of memory practices from western Europe to the rest of the world. In memory studies there is a need then to follow the lead of Chakrabarty (2000) and "deprovincialize Europe". The 1965 case is linked to multiple human rights regions of memory. The parallels that these links have in other countries, which have similarly decolonized during the Cold War, make it an excellent reference point for understanding global memory practices. In order to understand the regions of memory to which Indonesia is connected, however, some reflection on Indonesian history is needed.

The 1965 Genocide

The 1965 genocide in Indonesia occurred in the global context of the Cold War. The Dutch ruled over the colony of the Netherlands East Indies after the demise of the Dutch trading company from the eighteenth century until the arrival of the Japanese in 1942, who took over the colony during the Pacific War. Indonesians were then forced to fight in a war of independence with the Dutch from 1945 to 1949. Despite eventual Dutch recognition of Indonesian sovereignty in 1949, economic decolonization was far from complete, and the Netherlands and other countries continued to exert influence over the Indonesian economy into the 1950s and 1960s. Indonesia's ongoing decolonization coincided with the beginning of the Cold War, such that its domestic politics took on global significance in terms of a broader process of competition for influence between the Soviet Union, the US, and increasingly China, across Asia, Africa, and Latin America (Westad 2012). Prior to 1965, the Indonesian Communist Party (Partai Komunis Indonesia, PKI) enjoyed the patronage of the radical nationalist President Sukarno. There were heightened tensions between anti-communist forces in the army and the PKI by 1965. In this context, the United States also tried to support anti-communist factions in the army through the provision of military aid and training (Simpson 2008, 70–82).

On September 30, 1965, an armed movement calling itself the September 30th Movement kidnapped and killed six military generals and one lieutenant. The army, under the command of Major General Suharto, quickly suppressed the Movement and blamed the action on the PKI. In order to encourage anti-communist hysteria and to pave the way for a large-scale violent repression, the army began to spread propaganda about the Movement, emphasizing communist barbarity, as "evidenced" by the alleged mutilation of the aforesaid generals' corpses (McGregor 2007, 69). The army used this propaganda in order to frame the PKI as a national threat.

Between 1965 and 1968, the army and anti-communist groups killed approximately half a million largely unarmed Indonesians and imprisoned hundreds of thousands of people for varying periods of time without trial in prisons and penal colonies (McGregor et al. 2018, 12). The army justified their violent actions against the PKI as a campaign to crush national traitors. The targets of the repression were members of the PKI and affiliated or closely associated groups including organizations for youth, women, farmers, teachers, and university staff. By 1965, the Party had over 3 million members, along with 20 million members of affiliated organizations (Mortimer 1974, 366). Many Indonesian students studying especially in Eastern European countries were subject to screening by embassies and faced the choice of either returning home to be imprisoned or becoming political exiles.

Primary responsibility for the repression lies in Indonesian hands, but foreign governments also played a part. The US government supported a failed rebellion against the president in the 1950s, as well as numerous assassination attempts on his life into the 1960s (Kahin and Kahin 1995). US government officials encouraged the army to be ready to move against the PKI when a chance arose. Besides the US government, the governments of the United Kingdom, Australia, the Netherlands, and the Soviet Union either encouraged the violence or stood by as it unfolded (Schaefer and Wardaya 2013).

The violence of 1965 was thus not just a national affair. It was supported by broader forces and connected to global patterns of violent anti-communist repression of those on the political left. Foreign governments frequently encouraged developmentalist authoritarian regimes, even providing arms to support the military violence that underpinned such regimes, in return for access to markets (Westad 2012). The 1965 violence thus needs to be understood in the global context of the Cold War.

Failed Indonesian Efforts to Address this Case

Throughout the thirty-two years of the New Order regime (1966–1998) of President Suharto, the Indonesian army dominated politics and discourse, playing a combined military and political role. The repression of the PKI was celebrated as a victory over national traitors and over atheist and evil forces within the nation. Indonesians were reminded annually on the anniversary of the repression of September 30, 1965 to be forever vigilant with regards to the communist threat (McGregor 2007, 84–85). Meanwhile, former political prisoners and their families and the families of those killed in 1965 were subject to ongoing legal discrimination, community prejudice, and surveillance. In a climate of intense ongoing anti-communism, it was not possible to critically examine or protest against the violence.

Since the fall of Suharto in 1998, there has been greater freedom of expression and greater attention to human rights in Indonesia. This has created more space for advocacy on behalf of survivors of the violence, but there are ongoing sensitivities to this history often expressed in the form of intense anti-communism (McGregor and Fealy 2010; McGregor 2012). There has been some success in relation to increased recognition of survivors of the violence, but there have been many more setbacks and ongoing societal and government resistance to addressing this past. In the early post-Suharto period, human rights activists and even some progressive political leaders such as President Abdurrahman Wahid looked to transnational models of transitional justice for inspiration as to how Indonesia could settle outstanding human rights claims. Drawing on global memory, they considered the South African Truth and Reconciliation Committee (TRC) a model and began to work towards legislation for an Indonesian equivalent (McGregor and Setiawan 2019). By 2001, the National Commission of Human Rights and NGOs had prepared a draft bill for a TRC. Law no 27/2004 was passed to help set up a TRC, but due to a sticking point over proposed amnesties for perpetrators, the constitutional court cancelled the law in 2006 (Evanty and Pohlman 2018, 316–18).

One of the most significant achievements with regard to the 1965 case was the report of the National Commission of Human Rights into the 1965 violence. This report was the result of four years of detailed research across Indonesia commencing in 2008 and completed in July 2012. The report was to form the basis of evidence for an ad hoc Human Rights trial. The Commission found that gross human rights violations had taken place, including killings, exterminations, slavery, forced removal from an area, restrictions on physical freedom, torture, rape and other forms of sexual violence, forced disappearances, and persecution (KOMNASHAM 2012). The commission attributed responsibility for these crimes to several branches of the military: PANGKOPKAMTIB (Commander of the Operations Command to restore Order and Security), PANGDAM (Regional Military Commanders), and local military commanders. Consistent with human rights laws for how retrospective cases of human rights abuses should be handled formulated in the early post-Suharto period (Law No 26, 2000), the Commission recommended further investigations by Indonesia's Attorney General. The report also left open the possibility that the investigations form the basis for a non-judicial settlement if agreed to by survivors, such as a Truth and Reconciliation Commission. The Attorney General at that time, Basrief Arief, however, sat on the report until November 2012 before sending it back to the Commission asking for more witnesses, more information about the accused, and information on whether those who provided information had done so under oath (Waluyo 2012). This pattern of delay and obstruction and requests for more information continued, resulting in no action on the case and an embargo on the public release of the full report.

In anticipation of the 2014 presidential election, during which Prabowo, a former General, ran against Djoko Widodo, a businessman who was less well connected, Indonesian human rights activists escalated their efforts to draw attention to unresolved human rights cases in the hope this would be an election issue. They held a Year of Truth Telling, using public hearings to make known a range of unresolved human rights cases, including the 1965 case, and to increase understanding of the impact on victims of impunity such that presidential candidates would also pay attention to human rights (Pohlman 2016). These hearings were able to proceed, perhaps because they did not invoke a legal framing or focus on naming perpetrators. When the civilian Djoko Widodo was elected in late 2014, there were expectations that there would be greater attention to human rights, given his campaign declaration included a promise to respect human rights and uphold justice in the case of past human rights abuses. But whatever his original intentions, he quickly made deals with retired military politicians, including giving them key posts related to human rights cases, to strengthen his political position (McGregor and Setiawan 2019). This, combined with a lack of conviction on Widodo's part, led to little action being taken on human rights cases, and ongoing impunity and obstruction of justice-related efforts.

The People's Tribunal Aims and Format

The idea for the People's Tribunal first came about in 2013 during the long period in which justice efforts for the 1965 case seemed to have stalled. The trigger for this idea was a screening in the Netherlands of the now famous film *The Act of Killing*, in which former executioners based in the Sumatran city of Medan boast about killing people in 1965 and re-enact those killings, demonstrating macabre means of torture and execution. Highlighting the film's theme of enduring impunity, one former executioner called Zulkadry states that if people wanted to take him to the International Criminal Court (ICC) he would go freely, because he feels no guilt and is confident that "war crimes are always defined by the winners, like George W. Bush who got away with Guantanamo Bay" (Oppenheimer et al. 2012). Since its release in 2012, the same year as the National Commission of Human Rights' report, the film has begun to attract increasing global attention. Many people with no connections to Indonesia learned about this case of mass atrocity for the first time through the film and it received strong attention in film, human rights and academic circles. The same has been true of its sequel, *The Look of Silence*, which focuses on the brother of an executed person who interviews those in his community who had carried out the killings. In a sense then, following Erll's concept of travelling transcultural memory, the films demonstrate the potential transformation of the 1965 violence from a little-known case to one that was now making its mark in "global memory" (2011, 15). The responses to the films demonstrated the potential for new bases of solidarity for the victims of the violence, and prompted activists to think about using international pressure to push for faster action at home. This led a group of Indonesian activists to formulate the idea for a people's tribunal to be held outside of Indonesia (IPT 2016). It had to be outside Indonesia because of ongoing anti-communist pressure on events related to discussion of this past.

The reasons why activists chose an international people's tribunal as a way of drawing attention to this case are numerous. First, even if enough political support could be mustered, the ICC could not hear cases that occurred before 2002 and it was unlikely there would be enough support for a mechanism similar to Cambodia's very costly ECCC hybrid tribunals (Goldberg 2014). Second, international people's tribunals, like international war crimes trials, involve a sense of what Levy and Sznaider (2010, 19) term "public dramaturgy", which in turn leads to intense media interest, which is precisely what the organizers wanted and achieved. Third, like international trials, people's tribunals have broader educative and moral purposes. Although they are not legally binding, people's tribunals are intended as moral injunctions for cases where justice has not been served due to particular global constellations of power and global hierarchies of justice according to which subaltern groups, such as victims of an anti-communist repression, might rate lowly. The IPT 1965 organizers explained on the tribunal website:[1]

> IPT 1965 hopes to bring political and moral pressure to encourage people—namely, citizens, political parties, mass organizations, NGOs, victims and survivors, the international community, foreign states, UN agencies and other organizations—to urge the Indonesian government to conduct a formal judicial process; that is, to carry out thorough research, examine cases and testimonies of victims and survivors, and find a legal resolution.

The strategy of the group backing the tribunal was clearly to use international exposure to increase pressure on the Indonesian government to act to resolve the 1965 case. This is a strategy that Keck and Sikkink have referred to as the boomerang effect, by which they "bring pressure to bear from above on their governments to implement changes" (1998, 161). It is most often used when advocacy at home is unlikely to have an effect.

By using the term "crimes against humanity" in the title of the IPT 1965, the organizers sought to invoke international human rights norms, especially the United Nations Declaration of Human Rights, to which Indonesia is a signatory, as well as the legacies of the Nuremberg trials in which crimes against humanity were first tried. The use of this term, following KOMNASHAM's findings of crimes against humanity in the 1965 case, sought to put the Indonesian government on notice that crimes that are universally condemned had occurred in this case. The fact that the tribunal was held in the Hague, the same location as the ICC and the same city where the ICTY (International Criminal Court for the former Yugoslavia) had been held, further intensified a sense that Indonesian leaders, like the former leaders of Yugoslavia, could be held to account in an international forum.

Furthermore, the coordinator of the IPT 1965, human rights lawyer Nursyahbani Katjasungkana, had personal experience of people's tribunals through her involvement as lead prosecutor for the Indonesian team at the 2000 Women's International War Crime's Tribunal on Japan's Military Sexual Slavery (hereafter Women's Tribunal), held in Tokyo. In that tribunal she led the presentation of evidence from Indonesia, including testimony from Indonesian survivors of enforced military prostitution

[1] https://www.tribunal1965.org/en/faq (last accessed May 2022).

implemented by the Japanese military throughout the colony between 1942 and 1945 (WIWCT 2001, 4). A notable feature of the Women's Tribunal was the centrality of survivor testimony as a way of demonstrating the ongoing suffering of women both at the time of the violence and through post-war ongoing discrimination in their societies, as well as an emphasis on the need for punishment for the human rights crime of sexual violence (Sakamoto 2001, 52). The tribunal aimed to create a permanent public record of that violence, which was later preserved in a museum dedicated to the tribunal and related activism (Watanabe 2015, 237). Indonesian human rights activists thus drew on global memory and activist memory when thinking about how to increase the profile of the 1965 case. In this way, the IPT 1965 was connected to a human rights region of memory.

Like in the case of the Women's Tribunal, the organizers of the IPT 1965 sought to create a record of the violence from the perspective of survivors and based on extensive research by scholars and activists, particularly given that the full report of the National Commission of Human Rights into 1965 was not on public record. As General Coordinator, Katjasungkana explained that they wanted to provide "a public record of the mass killings and other crimes committed" (McGregor and Purdey 2015). In this sense, like the 2000 Women's Tribunal, survivor testimony and the attribution of blame for crimes against survivors were at the center of the tribunal. Due to budgetary and time constraints, only a select number of survivors spoke at the tribunal. Each survivor, however, represented a different kind of abuse. One anonymous witness testified to having found a mass grave in Wonosobo, Central Java, in 2000, relating to the mass killing (Muchtar 2015; see also McGregor 2012). Another anonymous witness from Eastern Indonesia testified that, based on her research into six regions, she had evidence that the killings were planned (Muchtar 2015). Martin Aleida, a writer imprisoned for a year, testified to being tortured during his detention in a Jakarta concentration camp (Wijanarko 2015). A seventy-year-old woman using a pseudonym gave a harrowing account behind a curtain of being sexually tortured by a university lecturer in Yogyakarta, Central Java, during purges of staff and students from university campuses (Kurniawan 2015).

In addition to the compelling testimony, the IPT 1965 also had performative dimensions. Like in the Women's Tribunal (Sakamoto 2001, 50), artwork by survivors was displayed on the walls in the foyer of the tribunal building the Nieuwe Kerk, an old Gothic style church in The Hague.

Participants lit candles for reflection and remembrance of those who had died in the killings or long after.

The main organizers of the IPT 1965 were the above-mentioned lawyer Nursyahbani Katjasungkana and the Dutch sociologist Saskia Wieringa, who acted as chair. There was an organizing committee and international steering committee consisting of 42 persons, 38 of whom were of Indonesian background, including members of the exile community in Europe (IPT 2016, 24–25). This included leading Indonesian human rights activists, lawyers and a team of researchers and journalists. It was supported in particular by the following NGOs which had worked on the 1965 case: Kontras (*Komisi untuk Orang Hilang dan Korban Tindak Kekerasan*, Commission for the Disappeared and Victims of Violence), AJAR (Asia Justice and Rights), and JPIT (*Jaringan Perempuan Indonesia Timur*, Network of Eastern Indonesian Women). It did not, however, have the support of all Indonesian human rights groups and campaigners, as some felt the tribunal was too provocative.

The Indonesian prosecution team consisted of five lawyers, headed by chief prosecutor Todung Mulya Lubis, a senior Indonesian human rights lawyer and former head of Indonesian Legal Aid (1980–1983). He was accompanied by Agung Wijaya, a human rights advocate, Sri Suparyati, Vice Director of Kontras, Antarini Arna, who was also a judge in the Women's Tribunal, and two other lawyers from Indonesian Legal Aid, Uli Parulian Sihombing and Bahrain Makmun (IPT 2016, 25). One representative from the National Commission of Human Rights (Dianto Bachriadi) attended, along with one member of the National Commission on Violence against Women (Mariana Aminuddin). These representatives provided briefings to the IPT 1965, confirming that the evidence produced by the tribunal accorded with the findings of their commissions' investigations. Responding to the testimony, scholarly submissions and information from the findings of reports from the two national commissions, the five-person prosecuting team prepared a legal analysis of all materials using international laws and conventions on human rights which was then submitted to the judges.

Through the selection of judges, the tribunal organizers also positioned the IPT 1965 in global struggles for justice and within a human rights region of memory. The Tribunal was overseen by Zak Jacoob, a supporter of the anti-apartheid movement who served as a South African constitutional court judge from 1998 to 2013 (Ferdianto 2015). Other prominent judges included QC Sir Geoffrey Nice, who had served on a panel of judges for the ICTY (International Criminal Court for the former Yugoslavia), Cees Flinterman, a former member of the high commission of the UN Human Rights Committee in Geneva, and Mireille Fanon Mendes, the child of Frantz Fanon, one of the most famous anti-colonialists, and a member of the Permanent People's Tribunal based in Rome (IPT 2016: 184–85). Three other activists with extensive human rights experience joined the panel: Shadi Sadr, Founder and Director of Justice for Iran, John Gittings, a retired journalist for *The Guardian* and expert on Cambodia, and Helen Jarvis, who worked on Cambodia's ECCC and who serves as Vice President of the Permanent People's Tribunal.

Although the findings of the tribunal were not legally binding, the inclusion of eminent jurists on the panel meant that the findings of these judges, like those in the case of the Women's Tribunal, could still be consulted as sources of international law (Ling 2009, 98–99). The accused party in the tribunal was the government of Indonesia. The government was on trial, and found guilty by the judges, of nine counts of human rights abuses, including killings, enforced disappearances, imprisonment, enslavement, torture, sexual violence, exile, hate propaganda, and genocide (IPT 2016, 171–72).

There were many misunderstandings in Indonesia about the status of the IPT 1965, and it seemed there was little awareness about people's tribunals and how they worked. To counter these misunderstandings and socialize the idea of the tribunal, its advocates set up a website with a facts and questions section. During the tribunal, the organizers live-streamed the proceedings via the IPT website. The Facebook group "Friends of the Tribunal" included posts of many photos of the tribunal and related media articles. For the duration of the tribunal from November 10 to 13, the Facebook page ran hot with posts.

The online content relating to the tribunal meant that even though many people affected by the violence or who remain committed to historical justice for this case, could not physically attend the tribunal in The Hague due to the prohibitive cost of airfares, they could still be virtual audience members. Organizers Katjasungkana and Wieringa (2019, 34) reported that tens of thousands of people watched the recordings on YouTube. In this sense the live-streaming of the tribunal. including screenings in Indonesia in public or in homes, facilitated greater possibilities of disseminating and globalizing memories of 1965.

Official Indonesian Responses to the Tribunal

In the months leading up to fiftieth anniversary of the beginning of the violence and the November tribunal, a series of events in Indonesia and abroad indicated there was growing sensitivity to the IPT 1965 and international attention to the 1965 case. In October 2015, Indonesia was the guest of honor at the Frankfurt book fair, under the theme of "the imagination of 17,000 islands". Because there were several new literary books on 1965, including a novel about political exiles called *Pulang* ("Home") by Leila Chudori, 1965 became an unintentional theme of the book fair. Coinciding with the half-centenary of the violence, this attracted German and further international attention. This resulted in domestic criticism of the Indonesian committee that had selected these books (Afrisia and Probo 2015).

In October 2015, the Indonesian police forced the Ubud Writers and Readers Festival, a huge regional arts event attracting thousands of Indonesian and foreign attendees, to cancel the screening of *The Look of Silence*, and a series of events dedicated to commemorating 1965 that I co-organized, including the launch of a translated series of Indonesian-authored books on 1965 (Kuddus 2017, 79–81). The cancellation again seemed to indicate increased nervousness about discussions concerning 1965.

In the months preceding the tribunal, the leading Indonesian lawyers representing the case were subjected to increasingly hostile attacks. Before the tribunal had even got underway, they were labeled "enemies of the state". In a heated discussion in a talk show on the day before September 30th hosted by the Indonesia Lawyers' Club, which featured IPT 1965 Coordinator Katjasungkana as a speaker, for example, retired General Kivlan Zain accused Indonesians of "reporting the government to the International Criminal Court" and of "being traitors to the country", stating that in 1965 that "we were only exacting revenge" on the PKI for killing the generals (Indonesia Lawyers Club 2015). His confusion of the IPT 1965, a non-legal mechanism, with the ICC is telling of military fears of international accountability for human rights crimes.

There was a clear atmosphere of tension around the tribunal. In his opening statement to the IPT, Chief Indonesian prosecutor Mulya Lubis disclosed:

> We come all the way from Indonesia not without risk. As human beings we are nervous and worried. We are worried because in our beloved country the matter related to 1965 atrocities is still taboo, not subject to discussion and deliberation. (Lubis 2015)

As anticipated, the tribunal hearings provoked extreme reactions in Indonesia. This included criticisms of the choice of location of the Netherlands, building on references to colonial violence and nationalist criticism of the activists behind the tribunal.

The Netherlands and Regions of Memory

Activists and survivors made the decision to hold the tribunal abroad because of repeated intimidations of conferences, events and justice initiatives related to 1965 in Indonesia, and because they sought to build international momentum on this issue. As reported on the IPT website, The Hague was chosen as a location because it is an international city of justice hosting not only the ICC, but also the Peace Palace, the site at which the first Hague Conventions were held in 1899, and because it was where the ICTY was held and where the judgement of the Women's Tribunal was delivered. In making this choice, the organizers attempted to invoke a region of memory that transcended nations and instead connected the 1965 case not with particular nations, but instead with many cases where there had been legal redress for human rights abuses.

Despite The Hague being the location of the tribunal, no role was played by the Dutch government. The Dutch government would presumably have little reason to support the tribunal given its representatives—following the US government—remained quiet about the 1965 killings, prioritizing economic ties over human rights. The former Dutch Minister for Development and Co-operation Jan Pronk, for example, stated in a seminar leading up to the tribunal in April that he was sympathetic to the tribunal in the context also of Dutch colonial atrocities against Indonesians, and because in his view, economic interests had always led the Netherlands government to remain quiet on the 1965 case (Santoso 2015).

Holding the tribunal in the Netherlands was, however, a calculated risk. Because the Netherlands is the former colonial power of Indonesia, there are ongoing sensitivities as to how the Dutch treated Indonesians in the past. These sensitivities have been heightened since 2011, with a series of successful court cases against the Dutch government for massacres of Indonesians during the independence war of 1945–1949, which brought memories of colonial violence and colonial ties to the fore in Indonesia. By the time of the IPT 1965, the Dutch government had apologized twice—in 2011 and 2013—and agreed to compensate surviving widows of the Rawagede and Westerling massacres which occurred during the independence war of 1945–1949. The Rawagede massacre took place in a village in West Java in 1947 and involved the killing of male villagers by Dutch colonial troops who were seeking out Republican soldiers (Scagliola 2012). The Westerling Affair of December 1946 was the most notorious Dutch atrocity during the independence struggle (McGregor 2014, 288–90). In this case, Dutch Special Forces under Captain Raymond Westerling's command killed thousands of civilians in South Sulawesi in an attempt to suppress opposition to the formation of the State of Eastern Indonesia.

In 2013, the Dutch government decided to apologize to the victims of all executions during the struggle (Santi 2013). It also offered compensation to surviving widows, though under strict conditions. As I have explained elsewhere (McGregor 2014, 295–98) the primary advocacy group behind these cases was a societal group the Committee for Dutch Honorary Debts, composed of Dutch and Indonesian activists. Although Indonesian politicians did not back these cases against the former colonial power with any vigor, some politicians immediately suggested in response to the IPT 1965 that the Dutch should settle their own records of violence against Indonesians first. The implication was they understood that the Dutch government was somehow involved in the tribunal.

The Co-ordinating Minister for Political Affairs, Law and Security, retired General Luhut Panjaitaan, issued a statement referring to Dutch colonial crimes and implying it was foreigners behind the trial. Panjaitaan said:

> now I am going to ask Westerling if he also wants to open up, how many Indonesians were killed? Don't just listen to the voices of white people (*bule*), listen also to the voices of Indonesians. (Franciska 2015)

In his mind, the tribunal was a foreign plot directed by white people, which, given most of the organizers were Indonesians, was untrue. By using this language and invoking the legacies of colonial violence, he cast this memory struggle between former colonizer and the former colonized, linking the two nations together in a region of memory.

There was a strong nationalist tone underpinning many of these responses. The Attorney General Prasetyo, who had notably taken no action on the 1965 case, similarly stated: "these are our own problems, and we'll solve them ourselves, there is no need for involvement by other parties" (Akuntono 2015). The immediate reference point was the Indonesian nation and a sense Indonesia needed to settle its own matters without foreign intervention. Yet it was precisely the intention of the tribunal to call out the Indonesian government for its failure to resolve the 1965 case. As one activist Aan Anshori (2015), from Kontras, noted: "Those active in the field of advocacy certainly understand the significance of shaming and naming in advocating for a case." The Jakarta secretariat head, academic Sri Lestari Wahyuningroem (2019, 189–92), has similarly argued that naming and shaming perpetrators, by which she presumably meant specific branches of the military and government, for responsibility for the 1965 violence, was a deliberate strategy used in the tribunal to put pressure on the government. The tribunal therefore sought to expose those responsible for the violence and the crimes committed, in the hope the government would finally address the case.

Despite the assertion of some that this was a national matter for Indonesians to deal with, a significant section of the tribunal addressed evidence of foreign complicity in the 1965 case, especially evidence of US, UK and Australian complicity. This is where another region of memory was evoked, the ties binding countries that supported anti-communist violence and human rights abuses during the Cold War and the so-called Cold War client states such as Indonesia under military rule. US foreign relations historian Bradley Simpson (2015) testified at the tribunal as an expert witness, stating that the US and other western countries had encouraged the army to strike against the communists at any opportunity. The American government had provided arms, finances and political support for the violence and was fully aware of the scale of the killings as they unfolded. The Australian and British governments conducted secret propaganda campaigns to assist the Indonesian army (see also Simpson 2008, 178–212).

What was the purpose of including foreign powers amongst those accused of responsibility? I believe the main reason was to stimulate more international awareness and actions from these parties. As noted above, until the Oppenheimer films, despite ongoing small-scale activism on this issue from political exiles and small activist groups based in the UK, Netherlands, US, Australia, and Canada, global awareness of this case was very low. By holding other states also accountable, the IPT 1965 organizers attempted to draw more attention to the case and spread the sense of responsibility for this injustice, so that differently implicated parties might begin to address this past triggering action from others. To date, however, no foreign government has formally responded to the tribunal's findings.

Conclusions

In terms of meetings its aims, IPT 1965 did increase pressure on the Indonesian government. It was only in the face of ongoing pressure from human rights groups, including preparations for the tribunal, that in May 2015 the Attorney General recommended an alternative, but flawed, initiative of a reconciliation committee to deal with all human rights cases (McGregor and Setiawan 2019). Further to this, in April 2016, Coordinating Minister for Law, Political Affairs and Security Luhut Panjaitan backed a national symposium on the 1965 violence called "Dissecting the 1965 Tragedy: A Historical Approach." This was the first gathering of this kind with government support (McGregor 2017). Although both these measures encompassed no legal forms of redress and implied the case could be dealt with purely by means of discussion, the tribunal organizers have through related publications, including the judgement (IPT 2016), created a publicly available record of crimes of humanity in this case. The resulting judgement furthermore provides legal arguments that may form the basis of future legal cases on this issue.

Through my analysis of the 2015 International People's Tribunal on Crimes Against Humanity in Indonesia 1965, I have demonstrated that in the face of persistent obstruction, Indonesian activists have constantly drawn on global models of justice for inspiration to make the 1965 case more visible globally. The tribunal and debates surrounding it demonstrate how the 1965 case is connected to several human rights regions of memory, due to historical experiences and parallel struggles for historical justice. These regions are not necessarily demarcated by geography and instead encompass connections resulting from broad processes such as colonialism, anti-communist Cold War alliances, or connections fostered with countries or human rights because of their transnational symbolism in international human rights struggles. They demonstrate the multiple and complex ways in which one human rights memory struggle may be bounded to many others.

References

Afrisia, Rizky Sekar, and Vega Probo. 2015. Memanfaatkan 1965 sebagai Tema Frankfurt Book Fair 2015. *CNN Indonesia*, July 1. http://www.cnnindonesia.com/hiburan/20150701143743-241-63612/memanfaatkan-1965-sebagai-tema-frankfurt-book-fair-2015/.

Akuntono, Indra. 2015. Jaksa Agung: Masalah HAM Kita Selesaikan Sendiri. *Kompas*, November 10. https://nasional.kompas.com/read/2015/11/10/10290171/Jaksa.Agung.Kasus.HAM.Kita.Selesaikan.Sendiri.

Anshori, Aan. 2015. IPT 65: Diantara Kepungan 4 Kelompok. *Kontras Surabaya*, June 30. http://kontrassurabaya.org/diskursus/ipt-65-diantara-kepungan-4-kelompok/.

Chakrabarty, Dipesh. 2000. *Provincializing Europe: Postcolonial Thought and Historical Difference*. Princeton: Princeton University Press.

Erll, Astrid. 2011. Travelling Memory. *Parallax* 17, no. 4: 4–18.

Evanty, Eva, and Annie Pohlman. 2018. After 1965: Legal Matters for Justice. In *The Indonesian Genocide: Causes, Dynamics and Legacies*, ed. Katharine McGregor, Jess Melvin and Annie Pohlman, 311–44. Cham: Palgrave Macmillan.

Franciska, Christine. 2015. #Tren Sosial: Salah Kaprah Soal Pengadilan Rakyat 1965 di Belanda. *BBC Indonesia*, November 11. https://www.bbc.com/indonesia/majalah/2015/11/151111_trensosial_tentang_ipt.

Ferdianto, Riky. 2015. Pengadilan Rakyat Kasus 1965 ini Hakim Yang Adilii Indonesia. *Tempo*, November 6. https://nasional.tempo.co/read/716542/pengadilan-rakyat-kasus-1965-ini-hakim-yang-adili-indonesia.

Goldberg, Leon. 2014. Human Rights Wonks Discuss The Act of Killing. *UN Dispatch*, March 1. https://www.undispatch.com/human-rights-wonks-discuss-the-act-of-killing/.

Indonesia Lawyers Club. 2015. 50 Tahun G30S PKI Perlukah Negara Minta Maaf. September 29. https://www.youtube.com/watch?v=V3iS-myzmfg.

IPT 1965. 2016. *Final Report of the IPT 1965: Findings and Documents of the International People's Tribunal on Crimes Against Humanity Indonesia 1965*. The Hague, Jakarta: IPT 1965 Foundation.

Hutwan, Arif. 2015. Ryamizardi Sebut Aktivis Pengadilan Rakyat Musuh Negara. *Metrotv News*, November 12. https://www.medcom.id/nasional/hukum/3NO4X30b-ryamizard-sebut-aktivis-pengadilan-rakyat-musuh-negara.

Kahin, Audrey, and George Kahin. 1995. *Subversion as Foreign Policy*. New York: New Press.

Katjasungkana, Nursyahbani, and Saskia Wieringa. 2019. Organisation and Impact of the International People's Tribunal People's Tribunal on 1965 Crimes Against Humanity in Indonesia. In *The International People's Tribunal for 1965 and the Indonesian Genocide*, ed. Saskia E. Wieringa, Jess Melvin and Annie Pohlman, 22–43. London: Routledge.

Keck, Margaret, and Kathryn Sikkink. 1998. *Activists Beyond Borders: Advocacy Networks in International Politics*. Ithaca: Cornell University Press.

Kimura, Ehito. 2015. The Struggle for Justice and Reconciliation in Post-Suharto Indonesia. *Southeast Asian Studies* 4, no. 1: 73–93.

KOMNASHAM. 2012. *Pernyataan Komnas HAM Tentang Hasil Penyelidikan Pelanggaran HAM Yang Berat Tahun 1965–1966*. Jakarta: Komnas HAM.

Kolimon, Mery, Lily Wetangterah and Karen Campbell-Nelson. 2015. *Forbidden Memories: Women's Experiences of the 1965 Violence in Eastern Indonesia*. Clayton: Monash University Publishing.

Kuddus, Rohana. 2017. The Ghosts of 1965. Politics and Memory in Indonesia. *New Left Review* 104: 45–92.

Kurniawan, Hasan. 2015. Saksi IPT Sebut-Maha-Guru-UGM-Terlibat-Penyiksaan. *Sindonews.com*, November 12. http://daerah.sindonews.com/read/1060978/189/saksi-ipt-sebut-maha-guru-ugm-terlibat-penyiksaan-1965-1447313839.

Levy, Daniel, and Natan Sznaider. 2010. *Human Rights and Memory*. Pennsylvania: Pennsylvania State University Press.

Ling, Cheah Wui. 2009. Walking the Long Road in Solidarity and Hope: A Case Study of the "Comfort Women" Movement's Deployment of Human Rights Discourse. *Harvard Human Rights Journal* 22, no. 1: 63–108.

Lubis, Todung Mulya. 2015. Opening Statement Chief Prosecutor of the International People's Tribunal 1965, 17 November 2015, https://www.tribunal1965.org/en/1965-tribunal-hearings-opening-statement-chief-prosecutor-todung-mulyalubis/.

Marching, Soe Tjen. 2016. Interview with an Activist, Soe Tjen Marching. *Inside Indonesia* 126, July–September. https://www.insideindonesia.org/interview-with-an-activist-soe-tjen-marching.

Mortimer, Rex. 1974. *Indonesian Communism Under Sukarno: Ideology and Politics*. Ithaca: Cornell University Press.

Muchtar, Bari. 2015. Berbagi Saksi Muncul di Hari Pertama IPT 1965 Den Haag. *Berita Belanda*, November 11. http://beritabelanda.com/berbagai-saksi-muncul-di-hari-pertama-ipt-1965-den-haag/.

McGregor, Katharine. 2007. *History in Uniform: Military Ideology and the Construction of Indonesia's Past*. Singapore: SEAP Series Singapore University Press, University of Hawaii Press and KITLV.

———. 2012. Mass Graves and Memory of the 1965–66 Killings. In *The Contours of Mass Violence in Indonesia, 1965–1968*, ed. Douglas Kammen and Katharine McGregor, 234–62. Singapore: SEAP Series Singapore University Press, University of Hawaii Press and KITLV.

———. 2014. From National Sacrifice to Compensation Claims: Changing Indonesian Representations of the Captain Westerling Massacres in South Sulawesi (1946–1947). In *Colonial Counterinsurgency and Mass Violence: The Dutch Empire in Indonesia*, ed. Bart Luttikhius and Dirk Moses, 282–307. London: Taylor and Francis.

———. 2017. Exposing Impunity: Memory and Human Rights Activism in Indonesia and Argentina. *Journal of Genocide Research* 19, no. 4: 551–73.

———. 2018. Historical Justice and the Case of the 1965 Killings. In *Routledge Handbook of Contemporary Indonesia*, ed. Robert Hefner, 129–39. London: Routledge.

McGregor, Katharine, and Greg Fealy. 2010. Nahdlatul Ulama and the Killings of 1965–66: Religion, Politics and Remembrance. *Indonesia* 89 (April): 37–61.

McGregor, Katharine, Jess Melvin, and Annie Pohlman. 2018. New Interpretations of the Causes, Dynamics and Legacies of the Indonesian Genocide. In *The Indonesian Genocide: Causes, Dynamics and Legacies*, ed. Katharine McGregor, Jess Melvin and Annie Pohlman, 1–26. Cham: Palgrave Macmillan.

McGregor, Katharine, and Jemma Purdey. 2015. Indonesia on Trial: Rights Activists Face Backlash. *Indonesia@Melbourne*, December 18. https://indonesiaatmelbourne.unimelb.edu.au/international-peoples-tribunal-activists-face-backlash/.

———. 2016. The IPT is an Historic Intervention. Will it Finally Lead to Action? *Indonesia@Melbourne*, July 29. https://indonesiaatmelbourne.unimelb.edu.au/the-ipt-1965-is-a-historic-moral-intervention-will-it-finally-lead-to-action/.

McGregor, Katharine, and Ken Setiawan. 2019. Shifting from International to "Indonesian" Justice Measures: Two Decades of Addressing Past Human Rights Violations. *Journal of Contemporary Asia* 49, no. 5: 837-61.

Oppenheimer, Joshua, C. Cynn and Anonymous. 2012. *The Act of Killing*. Denmark: Final Cut for Real.

Phillips, Kendall R., and G.M. Reyes. 2011. *Global Memoryscapes: Contesting Remembrance in a Global Age*. Alabama: University of Alabama Press.

Pohlman, Annie. 2016. A Year of Truth and the Possibilities for Reconciliation in Indonesia. *Genocide Studies and Prevention* 10, no. 1: 60–78.

Rothberg, Michael. 2009. *Multidirectional Memory: Remembering the Holocaust in the Age of Decolonization*. Stanford: Stanford University Press.

Sakamoto, Rumi. 2001. The Women's International War Crimes Tribunal on Japan's Military Sexual Slavery: A Legal and Feminist Approach to the "Comfort Women" Issue. *New Zealand Journal of Asian Studies* 3 (1): 49–58.

Santi, Natalia. 2013. Belanda Minta Maafatas Pembantaian Westerling. *Tempo*, September 12. http://www.tempo.co/read/news/2013/09/12/117512705/Belanda-Minta-Maaf-Atas-Pembantaian-Westerling.

Santoso, Aboedprijadi. 2015. The 1965 Controversy and Need for People's Tribunal. *Jakarta Post*, April 16. https://www.thejakartapost.com/news/2015/04/16/the-1965-controversy-and-need-people-s-tribunal.html.

Santoso, Aboedprijadi, and Gerry van Klinken. 2017. Genocide Finally Enters Public Discourse: The International People's Tribunal 1965. *Journal of Genocide Research* 22: 594–608.

Scagliola, Stef. 2012. Cleo's "Unfinished Business": Coming to Terms with Dutch War Crimes in Indonesia's Independence War. *Journal of Genocide Research* 14 (3–4): 419–39.

Schaefer, Bernd, and Baskara T. Wardaya, eds. 2013. *1965 Indonesia and the World*. Jakarta: Goethe Institut and PT Gramedia Pustaka Utama.

Simpson, Bradley R. 2008. *Economists with Guns: Authoritarian Development and US-Indonesian Relations, 1960–1968*. Stanford: Stanford University Press.

———. 2015. Senjata dan Dana Untuk Suharto dalam Pembasmi PKI. Testimony at the IPT, IPT website. November 13. http://1965tribunal.org/id/senjata-dan-dana-amerika-untuk-soeharto-dalam-membasmi-pki.

Waluyo, Andylala. 2012. Korban Peristiwa 1965–1966 Kecewa dengan Kejaksaan Agung. *Voice of Asia, Indonesia*, November 12. http://www.voaindonesia.com/content/korban-peristiwa-1965-kecewa-dengan-kejaksaan-agung/1543843.html.

Wahyuningroem, Sri Lestari. 2019. What's in a Name: Naming and Shaming in the Indonesian 1965 Mass Violence Discourse and IPT. In *The International People's Tribunal for 1965 and the Indonesian Genocide*, ed. Saskia E. Wieringa, Jess Melvin and Annie Pohlman, 179–97. London: Routledge.

Watanabe, Mina. 2015. Passing on the History of "Comfort Women": The Experiences of a Women's Museum in Japan. *Journal of Peace Education* 12, no. 3: 236–46.

Wieringa, Saskia E., Jess Melvin and Annie Pohlman, eds. 2019. *The International People's Tribunal for 1965 and the Indonesian Genocide*. London: Routledge.

Westad, Odd Arne. 2012. *The Global Cold War*. Cambridge: Cambridge University Press.

Wijanarko, Bagus. 2015. Kisah Martin Aleida yang Bakal Bersaski di Sidang 1965. *CNN Indonesia*, November 10. http://www.cnnindonesia.com/nasional/20151110123059-12-90674/kisah-martin-aleida-yang-bakal-bersaksi-di-sidang-1965/.

Women's International War Crimes Tribunal Judgement for the Trial of Japan's Military Sexual Slavery (WIWCT). vawwrac.org/judgement_e01.pdf.

The Legacy of Empire in East-Central Europe: Fractured Nations and Divided Loyalties

Krishan Kumar

A Region of Empires

East-Central Europe has been a region of imperial experience, if not actual imperial construction, for at least a millennium. It has been crisscrossed by the Holy Roman Empire, the Habsburg, Ottoman, Romanov and Hohenzollern Empires. In the interwar period of the 1920s and 1930s, short-lived independence gave way to the Nazi Empire. After 1945 there was the Soviet Empire. No other region of Europe, with the possible exception of the Balkans, has been so marked and shaped by empires. It is impossible to imagine that so long and profound an experience should not have left its mark, as much as deeply emotional childhood experiences leave their marks on us as adults. It may be putting it too strongly to say

K. Kumar (✉)
Sociology Department, University of Virginia, Charlottesville, VA, USA
e-mail: kk2d@virginia.edu

S. Lewis et al. (eds.), *Regions of Memory*, Palgrave Macmillan Memory Studies, https://doi.org/10.1007/978-3-030-93705-8_4

that empire is inscribed in the DNA of East-Central Europe. But it is not hard to find the signs of their persisting legacy throughout the region.[1]

In his famous essay of 1984, "The Tragedy of Central Europe", the Czech novelist Milan Kundera claimed that Central European countries such as Czechoslovakia, Hungary, and Poland had been "kidnapped" by their powerful neighbor to the east, the Soviet Union.[2] Though really part of the West, they had been captured by the East. "What does Europe mean to a Hungarian, a Czech, a Pole? For a thousand years their nations have belonged to the part of Europe rooted in Roman Christianity. They have participated in every period of its history. For them, the word 'Europe' does not represent a phenomenon of geography but a spiritual notion synonymous with the word 'West'" (Kundera 1984a, 33). The 1989 anti-communist revolutions in East-Central Europe and the accession of most of the countries of the region to the European Union must seem a fulfillment of the dreams and hopes of all those who thought like Kundera in the 1970s and 1980s.

[1] My grateful thanks to Chris Hann and Ildikó Bellér-Hann for their help with many aspects of this chapter. Thanks also to the editors of this volume for their many suggestions, which have been invaluable in the revision of the chapter.

In a recent volume on "liberal nationalisms and the legacy of empires", the editor Iván Dénes notes, with respect to Central and Eastern Europe, the need to investigate the "damnosa hereditas" of the empires in the region. "In order to understand [the region's problems] one needs to break with the traditional paths of national histories, by digging out the suppressed heritage of the Habsburg, Russian and Ottoman empires, including their dissolutions" (Dénes 2006, 8). It is a measure of the difficulties, and of how much still needs to be done, that not one of the contributors to the Central and Eastern sections of the volume chooses to engage with this aspect of the question, focusing instead entirely on nineteenth-century issues.

[2] The use by Kundera and others, such as Georgy Konrad, of the term "Central Europe", as opposed to "East-Central Europe", or the more cumbersome "Central and Eastern Europe", the terms more favored in academic discourse, is meant to indicate a distinct sensibility, and a distinctive set of values, that have traditionally been associated with the concept of "Central Europe" in literature, history, and philosophy. It also emphasizes the importance of certain countries—Czechoslovakia, Hungary, and Poland especially—which are said to have embodied those values most completely. These are also the "Habsburg" countries, which is why the Habsburg Empire figures so prominently in their thinking. Other countries in the region—Ukraine, Belarus, the Baltic countries—are not generally included in "Central Europe". Of course both "Central Europe" and "East-Central Europe" are designed to exclude Russia, hence also they are the preferred alternatives to "Eastern Europe", which before the fall of the Soviet Union encompassed all of the countries in the Soviet bloc. For a discussion of these concepts with bibliographical references to the relevant literature, see Kumar (2001, 74–103, 279–91); see also Okey (1992), Trenscényi (2017), Schenk (2017).

The divide between "Central" and "Eastern" Europe is questionable, as is the whole attempt by many Central European intellectuals to exclude Russia from "Europe" (Kumar 2001, 71–103). But it is interesting that, in making his case for a distinctive Central European identity, Kundera looked back with much fondness to the Habsburg Empire and its role as the protector of the Central European nations against the designs of both of its powerful neighbors, Germany to the West and Russia to the East. Here he explicitly referred back to the "Austro-Slav" program of František Palacký and others, and their dream of Central Europe as "a family of nations" under the all-embracing protection of the Habsburg Empire. "The Austrian empire had the great opportunity of making Central Europe into a strong unified state. But the Austrians ... did not succeed in building a federation of equal nations, and their failure has been the misfortune of the whole of Europe. Dissatisfied, the other nations of Central Europe blew apart their empire in 1918, without realizing that, in spite of its inadequacies, it was irreplaceable" (Kundera 1984a, 33–34; see also Johnson 2002, 174–75).

The Habsburg Empire has been the object of much nostalgia in the region, not only in Austria but in several parts of the old Austro-Hungarian lands, including Galicia, Slovenia and Croatia (Kann 1980; Boym 2001, 219–47; Beller 2006, 195, 282–93; Hann 2012, 6; Kożuchowski 2013). That memory, over-rosy as it might be, is itself an important component of the legacy of empire in the region. But one must beware of the tendency, apparent in the case of Kundera as of other Central European intellectuals, to narrow the historical perspective to the Habsburgs, and to ignore, belittle, or marginalize the other empires that played so important a part in the region. The Ottomans, after all, ruled most of Hungary for more than a hundred and fifty years; the Prussians/Germans were an important presence in Poland for centuries; the Russians cannot be left out of the Baltic countries, nor Poland, Ukraine, Belarus. The Habsburgs might have become peculiarly identified with "Central Europe", but if we think in particular of the wider region of "East-Central Europe", the overwhelming fact is the play of many empires.

In choosing "East-Central Europe" as a more capacious, and historically more meaningful concept, one has to admit that, like "Central Europe", it contains its own ambiguities and its own biases. Besides the "core" countries—former Czechoslovakia, Hungary, Poland—contained in the Habsburgian "Central Europe" concept, it includes countries to the east and north of these, such as the Ukraine and the Baltic countries—Lithuania, Latvia, Estonia. It can also sometimes include some of the Balkan countries, such as Slovenia, Croatia, and Romania, that were formerly part of the Habsburg Empire. What it shares principally with "Central Europe" is the exclusion of Russia, together with an awareness of its looming presence over many centuries. Russia is not a part of either Central or East-Central Europe; yet in some way, at least partly, it defines those regions.

In view of the negativity towards Russia expressed by many Central Europeans, and their tendency to deny it any real or lasting influence, it is particularly important to stress its enormous significance in the region.[3] When in 1980 Hugh Seton-Watson edited a special issue of the *Journal of Contemporary History* on "Imperial Hangovers", he commented that while most of the European empires had disappeared following the two World Wars, the same was not true of Russia. Under the guise of the Soviet Union, it was still extant and consequently there was no entry on Russia (Seton-Watson 1980, 4).

The Soviet Union followed the other empires in 1991 and it has certainly had its "imperial hangover". But still it is important to remember how long the Russian imperial presence remained in the region. It dates back to at least the seventeenth century with the still controversial "agreement" of the Ukrainian Cossacks under Bohdan Khmelnytsky in 1654 to accept incorporation into the Russian Empire (Plokhy 2015, 97–107). It continued with the partitions of Poland in the eighteenth century and the conquest of the Baltic lands. The Russian footprint in the region is huge, and unmistakable. Moreover, unlike the Habsburg Empire, its imprint is still very much in evidence. Russian designs in the region even after the fall of the Soviet Union remain very obvious to Ukrainians, Belarusians and others, as Russian annexation of the Crimea well illustrates.

From Empire to Nation State

The European empires broke up in two waves, one after 1918 (the Continental land empires), the other after 1945 (the overseas empires). What emerged from them was a host of nation states, many of them newly

[3] It was strikingly evident to me, lecturing to graduate students from the region at the Central European University in Prague in the 1990s, how indifferent they were to Russian culture and history, and how strongly they insisted that they were even beginning to forget the Russian language that most of them had had to learn at school. It is part of the misfortune of the recent history of the region that Russian culture, once admired there, has been so comprehensively rejected, to the region's impoverishment (while there has been almost slavish adulation of Western thought and culture).

hatched. So striking was this development, and so speedy especially after 1945, that there was a tendency to regard the move from empire to nation state as somehow foreordained, at least in the modern world. Empires—multinational, dynastic—were out, nation states, embodying the people or nation, were in. Indeed, as exemplified by the United Nations—more properly, a league of nation states—they seemed to be the only internationally recognized and legitimated form of political organization.

Central-Eastern Europe was no exception. On the contrary, it virtually supplied the rule, in the sense that after the First World War it was there that the first great expression of the phenomenon took place and was most widely discussed. In line with President Woodrow Wilson's principle of self-determination and under his energetic promotion, the Habsburg, Ottoman, Hohenzollern, and Romanov Empires yielded up their successor nation states.[4] Those in the Russian orbit—Ukraine, Belarus, Georgia, Armenia, Azerbaijan and others—were rapidly re-absorbed in the new Soviet empire. But Poland fought off the Red Army and established a large independent state that included most of the old Austrian Galicia as well as sections of Lithuania, German Posnania and Silesia, and western Ukraine. Out of the Habsburg Empire came not just small Austria and a greatly diminished Hungary, but also Czechoslovakia, Slovenia, Croatia, and Bosnia. These last three, together with the other Western Balkan states that had emerged from the Ottoman Empire in the late nineteenth and early twentieth centuries, soon formed into the federation of Yugoslavia. In 1918 the Ottoman Empire was itself reduced to its core, the new nation state of Turkey.

The new states of East-Central Europe were creations of the imperial powers that dominated the Versailles settlement: the United States, Britain, France. Imperial thinking and concerns played a large part in that settlement and in the League of Nations that came out of it (as of the United Nations later), as Mark Mazower (2009a) has shown.[5] The intention was not simply to contain (and punish) Germany, but also to set up a chain of buffer states, overseen by the League of Nations, between it and

[4] Erez Manela (2007) discusses Wilson's principle of self-determination and vividly shows its worldwide impact, especially on the non-Western world. For the fall-out in Central and Eastern Europe, see Magocsi (1995, 125–29), Johnson (2002, 171–96), Roshwald (2001).

[5] The Mandates Division of the League of Nations played a particularly important role in seeing that the interests of the imperial powers were represented, though as with the later United Nations it was not always possible to contain anti-imperial sentiment, see Pedersen (2015).

the Soviet Union. The League of Nations, in other words, was to perform the task of the old Habsburg Empire; it proved singularly unequal to the task, another reason—among many—for the Habsburg nostalgia of the interwar period (Kożuchowski 2013).

But what were the new states that emerged or re-emerged after 1918? What was their character? How far did they correspond to nation states? What nations did they enshrine? Certainly, there were historic states and historical memories that were freely and passionately drawn upon by nationalists in the region. There was the medieval Crown of St. Stephen that formed the basis of the national claims of the Magyars. There were the Crown Lands of the Kingdom of Bohemia that played the same role for the Czechs. Poland had the Kingdom of Poland, as well as the memories of the great Polish-Lithuanian Commonwealth.

These very claims, however, were fraught with conflicts and contradictions. Above all they flew in the face of nationalist principles. None of the old kingdoms originally had of course any shred of national character. They were dynastic monarchies, incorporating a host of different ethnicities and nationalities. When, in the nineteenth and early twentieth centuries, nationalists reached back to them to give historical respectability to their claims, it took a good deal of intellectual ingenuity to map the new nations on the old kingdoms. Of course, all nationalisms were faced with this problem, Germany and Italy no less than Poland and Hungary. But the incorporation of much of East-Central Europe in competing empires had led to a degree of mixing and arbitrary partitioning of populations unequalled in the rest of Europe. As Gale Stokes put it, "Eastern Europe is a shatterzone whose ethnic map in 1900, if indeed one could be drawn, would resemble the cracked bottom of a dried mud puddle" (1997, 8).

The new Czechoslovakia, for instance, included not just Czechs (and Moravians, historically linked to the Czechs), but large minorities of (Austrian) Germans, Slovaks, Magyars, and Carpatho-Ukrainians. Until 1939 the minorities outnumbered the Czechs (Kann 1980, 51 n. 16). Germans—and Jews—had been a presence in Bohemia and Moravia for centuries. But Slovaks had been part of the Kingdom of Hungary, not of Bohemia, for over a millennium. "Jan Hus was not a Slovak hero nor Bílá hora a Slovak tragedy. Saint Stephen meant more to most Slovaks than Saint Václav" (Sayer 2000, 170). As justification for the inclusion of Slovaks, some attempt was made after 1918 to summon up memories of the ninth-century Greater Moravian Empire, which included Czechs and Slovaks (Magocsi 1995, 10). That this attempt to conciliate the Slovaks failed—ultimately leading to the "velvet divorce" of 1992 between Czechs and Slovaks—should not lead us to dismiss such appeals to the distant past as trivial: they were the common currency of many nationalist claims in the nineteenth and twentieth centuries, often with significant consequences.

Newly independent Poland, which was the product of the collapse of the three empires—Russian, German, and Habsburg—that had divided it up in the eighteenth-century, was also fractured along ethnic lines. This was inevitable given the determination of both Roman Dmowski and Józef Piłsudski to re-establish the borders of the old Polish-Lithuanian Commonwealth, though they had different ideas as to its final political form. The fledgling state was consequently involved in wars with Germany, newly independent Lithuania, the newly declared Ukrainian Republic, and the Bolsheviks in their quest to recover the lost territories of the old tsarist empire. The defeat, under Piłsudski's masterly command, of the Bolshevik forces at the Battle of Warsaw in 1920 was the most important victory for the new state. It later became part of the national myth that Poland had "saved the West", not entirely implausible, given that the Red Army had hoped that it would march on Berlin after taking Warsaw and launch the "world revolution" from there (Davies 2011, 101; Johnson 2002, 189).

The Treaty of Riga (1921) with the Soviet Union did not give Poland all of the eastern regions of pre-partition Poland that it wanted, but it gave it a good part of the old Habsburg province of Galicia to the south, including its capital, Lemberg or Lwów (hence, despite its large Ukrainian population, denying it to Ukraine). In the face of bitter Lithuanian opposition, Poland also incorporated the historic Lithuanian capital, Vilnius—but by now a predominantly Polish city (Wilno) and a center of Polish culture—in the new Polish state.

But those victories were purchased at the cost of a legacy of intense inter-ethnic conflict. The "Second Polish Republic" was only two-thirds Polish (and Roman Catholic); the remaining third was made up of Ukrainians (fifteen per cent), Jews (nine per cent), Belarusians (five per cent), and Germans (two per cent). This was, from a nationalist point of view, better than the Polish-Lithuanian Commonwealth, which had been only half Polish and Roman Catholic (Johnson 2002, 190; Davies 2011, 103). But the difference was that the Commonwealth of course had no aspirations towards nationhood; the Republic on the contrary was created in the spirit of the resurrection of the Polish nation. In an era of high nationalism, the minorities—with the exception of the Jews, who had no nation—were bound to be a source of tension as they strove to join their "own" nations.[6]

Finally, the case of Hungary represents in a sense the opposite of Czechoslovakia and Poland. If the latter two had large minorities, the Hungary that emerged after 1918 had virtually none. The application of the principle of self-determination in this instance—it was never consistently applied—meant the dismemberment of the historic Kingdom of Hungary, joint-partner in the Austro-Hungarian Monarchy. By the Treaty of Trianon (1920) Hungary lost two-thirds of its territory and nearly two-thirds of its population. Instead, nearly a third of ethnic Hungarians now lived outside Hungary, mainly in Yugoslavia and Czechoslovakia. For Hungarians, this meant the destruction of a thousand-year history, the history of the Crown of St. Stephen. Even under the Habsburgs, they had been able to enjoy rule over a large territory and many peoples. Now they were shrunk to a small core in the Danube basin. This reversal was unacceptable to most Hungarians and they vowed to undo it. A popular prayer after the Treat of Trianon ran: "I believe in one God, one Fatherland, and the Resurrection of Hungary" (in Johnson 2002, 195).[7]

It is ironic that, in the wake of Versailles and Wilson's principle of self-determination, it was only the two centers of the former multinational Habsburg Empire—Austria and Hungary—that looked at all like "nation states", in the sense that they were both more or less ethnically homogeneous. Not that this was any source of comfort to either of them. Small Austria wanted to join its German co-nationals in a "Greater Germany" and had indeed voted to do so in 1918, but was forbidden by

[6] It can fairly be said that the Ukrainians and the Belarusians did not properly have their own nations either, simply in both cases Soviet republics that claimed to represent them. But that did not necessarily still nationalist claims and hopes to create "true" Ukrainian and Belarusian nation states. For a good examination of the Belarusian case, see Rudling (2015).

[7] Chris Hann notes that indignation at the "injustice of Trianon" has re-awakened in Hungary with the end of socialism after 1989, and the difficulty of relating to the authoritarianism of the Horthy regime of the interwar period and collaboration with the Nazis in the Second World War. What is left, in the face of this "inglorious history of the twentieth century", is reversion to the memory of the "Thousand-Year" Kingdom of St. Stephen, and especially Hungary's role as "equal partner" in the nineteenth-century Habsburg Empire. "It seems that Hungary is still struggling to come to terms with the loss of imperial power … Its contemporary leaders would like nothing more than to blot out the entire twentieth century and return to the nineteenth, when they constructed their splendid capital city and the key elements of the richly ambiguous mythomoteur that remains in place today" (Hann 2015b, 103, 118).

the Entente Powers (Hitler granted them their wish in the *Anschluss* of 1938).[8] Hungary, with over three million Magyars in the "near abroad", sought to restore the territory of the Kingdom of Hungary and Hungary's historic place in the region. As in the nineteenth century, here nationalism could look like, and take the form of, imperialism.

It was the veteran Austro-Marxist theoretician Karl Renner, first chancellor of the provisional republic of "German Austria" in 1918–1919, and later, after 1945, President of the Second Austrian Republic, who drew the parallel between the old empires and the new nation states formed at Versailles. But he also pointed to the significant difference, a difference that was fraught with dangers for the future.

> The former [Habsburg] Empire never pretended to be a national state, but the new national states were falsely proclaimed as such, and a large part of the domestic difficulties which beset them is due to this pretense. The peace treaties did not solve the problem of multi-national states but transferred it from each of the big powers to several small ones. (in Johnson 2002, 195; see also Kann 1980, 48)

In view of the common opposition of "nation state" to "empire", it is important to stress the extent to which the nation states that emerged after 1918 were the products of empire and bore its stamp. It was not only that it was the clash of empires in the First World War that had brought about the downfall of so many of the land empires—Ottoman, Romanov, Habsburg, Hohenzollern—and so allowed for the creation of the new nation states. Contrary to common claims, of both nationalists and many scholars, it was not nationalism that brought down those empires. Rather, it was other empires: principally the French, British, and American, with the Russian Empire also playing a role until the war between empires ended the rule of the Romanovs in 1917 (Kumar 2010, 136). In destroying each other, they created the space for the principle of national self-determination—in severely qualified form—to express itself. But the victorious powers meeting at Versailles made sure that they oversaw the

[8]Steven Beller (2006, 200) notes that "forbidding the union of Austrian Germans with Germany contravened the principle of national self-determination, and it had a disastrous effect on Central European politics". In that sense the new Austrian Republic formed in 1919 was not really a nation state at all but simply "what remained" of Habsburg Austria or "Cisleithania". Kann (1980, 42–43) notes that "in late 1918 and 1919 a sizeable majority of the Austrian people ... favoured Anschluss".

process in the interests of their own surviving empires. "From empire to nation state" was not some natural evolution but a carefully orchestrated and watchfully supervised operation by the empires themselves.

But the link between empire and nation state was even closer. The nationalisms that were the bases of the new nation states had in many cases themselves been the creations of empires, pursuing their own interests (Kumar 2017, 241–42, and *passim*). The Habsburgs stimulated Ukrainian nationalism in Galicia, as a rival force to the dominant Polish nationalism in the province; the Romanovs did the same with Belarusian and Ukrainian nationalism, also as a foil to Polish nationalism in Russian Poland. In Bohemia the Habsburgs found it politic to foster Czech nationalism against the stridency of a German nationalism calling for union with Germany. In Hungary the Habsburg rulers again found it helpful to cultivate the nationalism of the South Slav nations against the claims of Hungarian nationalism.

The empires not only stimulated nationalism, but they also often played a key role in the very forms of the nation states that succeeded them. The boundaries of most of the successor states to the Habsburg Empire were outlined in the federal scheme for the re-organization of the Austro-Hungarian Monarchy proposed by the Emperor Charles in 1917, following the death of Franz Joseph. The ethno-national units that were identified here were much as those to be found in the many such plans for reforming the Empire, especially those proposed by Austro-Slavists and Austro-Marxists in the late nineteenth and early twentieth centuries (Judson 2016, 420–33). The Empire had named, if not conceived, its heirs.

Norman Davies similarly has shown how it was the Germans who "laid the foundations on which Polish independence was later built" (2011, 98). It was Germany—pursuing its own interests, of course, as it did in sending Lenin in a sealed train back to St. Petersburg to create havoc in Russia—which restored the Kingdom of Poland in 1916 and set up a Polish Council of State together with twelve ministries. When the Germans withdrew from Warsaw in 1918, it was the German-appointed Regency Council that was left in control; and it was the Germans who, a week before withdrawing, released Piłsudski from prison in Magdeburg Castle and conveyed him to Warsaw, thus ensuring that it was he, not Dmowski—the favourite of the Polish National Committee in Paris—who would be the leader of the new Republic. "Poles", says Davies, "may be forgiven for believing that in 1918 they had 'fought their way to independence'. But that was not the case" (Davies 2011, 100).

Empire Upon Empire

"Imperial hangovers", or legacies, can come in layers. The new states of East-Central Europe were formed out of one wave of imperial dissolution after the First World War. Many of the problems of the interwar period carried the marks of the experience of end of empire, but also of what had preceded it, in the long life of the empires themselves. Struggles between Czechs and Slovaks, and between Croats and Serbs; the problem of the "Sudeten Germans"; the status of Vilnius/Wilno and of Lwów/Lviv and the conflict between Poles, Lithuanians, and Ukrainians; the fate of the Jews who had long enjoyed the protection of empires and had been among their most fervent supporters: all these were continuations in the new nation states, of matters that went back deep into the nineteenth century and earlier (Seton-Watson 1967; Crampton 1997, Part I; Snyder 2003, chs. 3 and 7). Imperial and post-imperial questions and concerns were linked by an umbilical cord.

The Second World War brought an extraordinary twist to this story. Not only did these historic rivalries bring about terrible and bloody conflicts, culminating in the effort to extinguish the Jews altogether (Snyder 2003, chs. 4 and 8; Snyder 2010). After 1945 there was first a second wave of imperial dissolution in the West. The British, French, Dutch, Belgians and finally Portuguese gave up, or were forced to give up, their overseas empires. At about the same time, however, there was also the re-imposition of empire in the East. After a life and death struggle with Nazi Germany, the Soviet Union consolidated its position in Eastern Europe. It re-absorbed the Baltic states lost in 1918: Polish-Lithuanian Vilnius/Wilno was now in the Soviet Union. It lopped off a large chunk of pre-war eastern Poland, including Lwóv/Lviv, adding it to an expanded Ukraine and Belarus. It established what was in effect an "informal empire" in East-Central Europe. Czechoslovakia, Hungary, Poland, East Germany, Bulgaria, Romania, and (until 1948) Yugoslavia found themselves part of the "Soviet bloc", which meant being client states of their powerful neighbor to the East (the analogy might be between the traditional Chinese Empire and its many tributary states).

This was a truly historic, not to say traumatic, reversal for the states of East-Central Europe. Nurtured, sometimes virtually being invented, in the bodies of empires over several centuries, they were given their freedom by the Entente Powers in 1918. This freedom was short-lived, lasting not much more than twenty years. After 1945, though most of them were formally independent, they were once more in the shadow of empire, linked by the Soviet-inspired doctrine of the "international solidarity of proletarian nations". That this was not an empty rhetorical figure of speech was shown time and again—1956, 1968, 1980–1981—as the Soviet Union intervened in its name to put down internal dissent in the nominally independent "Peoples' Democracies" of the Communist bloc.

There is no need to idealize the interwar period of independence to underline the significance of this reversal. Before 1939, authoritarian rule and flirtations with fascism developed in several of the East-Central European states, e.g., Austria, Hungary, Poland, Romania, Bulgaria, the Baltic states. Many collaborated with Nazi Germany—especially in the assault on the Jews—in the Second World War (this included Poles and Ukrainians in German-occupied territories). It is no surprise to learn of nostalgia for the old empires among several groups—Jews especially of course—faced with the consequences of national independence. The Austrian writer Joseph Roth, author of the famed novel of the Habsburg empire, *The Radetzky March* (1932), was only one of many artists and intellectuals who in the 1930s mourned its death (Beller 2006, 213–14). If empires had not been able to solve their problems, nationality and other, nation states seemed even less well-equipped to do so.

Nevertheless, the loss of independence and the fact that its time had been of such short duration mattered greatly in the psychology and mentality of the East-Central European states. It emphasized once more the longstanding condition of dependency, the fact that, as Milan Kundera put it, the peoples of Central Europe did not "make history", they were always "the object of history" (Kundera 1984b, 275; see also Kumar 2001, 16–27). Dependency, especially for many writers and artists in the region, bred a characteristic attitude of irony and passivity, a skepticism towards utopian ideas and hopes for betterment coupled with often wild swings of humor and hilarity. Franz Kafka's "there is plenty of hope, only not for us", seems emblematic of this Central European outlook; Jaroslav Hašek's *The Good Soldier Ŝvejk* (1921) is another typical expression with its exuberant satire, irreverence, and ethic of survival.[9]

[9] The full quote from Kafka is: "We are nihilistic thoughts that came into God's head … There is plenty of hope—no end of hope—only not for us!" (in Brod 1947, 75). The well-known joke, attributed to Galicians about the Galician front in 1914, is another good expression of some typical Central European attitudes. A German officer reports: "The situation is serious, but not hopeless". The Austrian officer retorts: "The situation is hopeless, but not serious" (Davies 2011, 468). Davies' chapter on Galicia gives a very good picture of a very

Of course, we should remember that "dependency" is not peculiar to East-Central Europe. Many nations have lived in a state of dependency for much of their history: Irish, Scots, and Welsh in a United Kingdom dominated by the English; Catalans and Basques in a Spain dominated by Castile; Quebecois in English-dominated Canada. Nor should we make too much of the fact that East-Central Europeans have often been ruled by "foreigners", e.g., Germans, Hungarians, Turks, Russians. Before the coming of nationalism, that was the rule rather than the exception in most European countries—and elsewhere—as a consideration of the ethnic roots of most of the European sovereigns before 1914 would quickly reveal.[10] For East-Central Europeans, as for most people before the coming of nationalism, what mattered most was region or locality, or perhaps religion, not nationality; and the idea that they should be ruled by their "fellow-nationals", not "foreigners", would have seemed strange to them (Gellner 1983, 8–18).

Nationalism did generate, among intellectuals at least, a demand for "self-rule", rule by members of one's own nation, however that entity was defined. That brought to the fore the consciousness, or at least the claim, that the empires were "prison-houses of nations", as it was said famously of the Habsburg Empire. But equally important to East-Central European intellectuals was the rise of Western liberalism, and the comparison between their condition and that of the peoples of Western European countries such as Britain and France. In those countries, and increasingly

characteristic part of East-Central Europe. See also Wolff (2010). Milan Kundera sees the "tragicomical" as the typical Central European outlook, a "view of the world which considers the comical side as an indivisible part of every human situation. Nobody, nobody is spared the comical which is part of our condition, our shadow, our relief and our condemnation (quoted Rupnik 1989, 204). Elsewhere he speaks of "our special humour: a humour capable of seeing history as grotesque" (1984b, 27). For another account of Central and Eastern European "mental reflexes", see Fodor (2013, 409–10). It is probably worth stressing that not all thinkers and artists from the region adopted this ironic or acerbic attitude.

[10] A good example would be the English—themselves invaders of the British Isles—who for long lived under French Normans (the Norman and Angevin kings), then under Welsh (the Tudors), then Scots (the Stuarts), then Germans (Hanoverians and Saxe-Coburgers), whose descendants still rule. Arguably the English have not governed themselves since Harold Godwinson was defeated by William the Conqueror in 1066. One could also mention the experience of India, under "foreign"—Muslim and British—rule from the eleventh to the twentieth centuries; or China, ruled by the Mongols (Yuan dynasty, thirteenth to fourteenth centuries), then later by the Manchus (the Qing dynasty, 1644–1912). Most states, for most of the time, have been ruled by "foreigners": that was virtually the norm before the nineteenth century.

in the West, they saw republicanism, parliamentarism, and the rise of representative democracy. It was this, even more perhaps than nationalism, that they saw stifled in the empires, with their traditions of authoritarianism and militarism (Dénes 2006). "Self-rule" therefore was opposed to dependency not only in the sense that nations should rule themselves, but also in the sense that the mass of people should rule themselves, that popular sovereignty should take the place of despotic and authoritarian rule.

For a brief period in the 1920s and 1930s, the countries of East-Central Europe had the opportunity to try out some of these ideas. The fact that they failed almost everywhere—Czechoslovakia is the one place that is generally held to have had some success with liberal democracy—led many, both within and outside the region, to conclude that East-Central Europe's long experience of empire had inured it to authoritarianism and made it incapable of genuine self-rule. Not only was there little experience of formal democracy; there had not even been the development of a strong civil society—a public sphere of non-governmental associations—within which people learn the norms and practices of public deliberation and collective decision-making. The empires had enfeebled citizenship and civic participation. That is why democracy failed in the interwar period.

The re-imposition of empire in the region after 1945, according to this view, only confirmed East-Central Europeans in this pattern of dependence. Public life was once more suppressed; any buds of civil society were cut off. The best one could do was to turn one's back on the regimes, to cultivate whatever private spaces one could find with friends and family. The rest was to conform, at least publicly. The emblem of this conformity is Vaclav Havel's famous greengrocer, who hangs in his shop window the sign, "workers of the world, unite!", not because he believes in it but because "it is one of the thousands of details that guarantee him a relatively tranquil life 'in harmony with society'". It carries the message "I am obedient and therefore have the right to be left in peace" (Havel 1987, 41–42). With infinitely greater power at its disposal, the Soviet empire confirmed and strengthened the depoliticization that had been instilled in the populations of East-Central Europe by the earlier centuries of empire.

This is a fairly well-aired view, which does not mean that it lacks plausibility. But two things need to be said about it. First, it makes the assumption that the imperial experience has been all of a piece and that there are not important variations within it. Secondly, it tends to assume that Soviet domination simply resumed the rule of empire that had been briefly interrupted by the period of independence between the wars. In both cases there is a certain flattening out or a homogenization of the impact of empire on East-Central Europe. We need to look more closely at the differentiated experience of empire in the region, especially if we want to make sense of some important contemporary developments.

Fault Lines and Fractures

Three empires in particular—the Ottoman, the Habsburg, and the tsarist Russian Empire—dominated the region, from the fifteenth to the twentieth centuries.[11] For all their similarities as dynastic, autocratic empires, they exhibited significant differences in the way they ruled their subject populations.

Like other Islamic empires, the Ottoman Empire was famously tolerant of other religious groups, Jews and Christians especially, whom it recognized as *dhimmi*, "people of the book". Religions, including Islam, were organized in self-governing *millets* (Kumar 2017, 97–109). This had important repercussions in the Ottoman's Balkan possessions, most of which had Christian Orthodox populations (the sizable Muslim population in the Balkans is another legacy, alongside the survival of Orthodoxy). Greeks in particular had a privileged position in the Empire and by the eighteenth century virtually became co-rulers with the Ottomans. Only with the rise of Greek nationalism and the Greek War of Independence did the decisive parting of the ways occur (with tragic consequences for both Greeks and Turks).

The Ottomans also ruled much of Hungary for over a hundred and fifty years from the defeat of the Hungarians at the Battle of Mohacs in 1526 to the Treaties of Karlowitz (1699) and Passarowitz (1718), when Hungary reverted to Habsburg rule. The Ottoman legacy in Hungary is complex and much disputed, though Turkish baths, Turkish mosques, some Turkish loan words and elements of Turkish dress and cuisine are standing reminders (Pálffy 2001, 124; Fodor 2013, 408). But among the most important effects of Ottoman rule in Hungary is the continuing strength of Protestantism there. When the Reformation occurred in Europe in the sixteenth century, the Spanish and Austrian Habsburgs took the lead in

[11] Germany of course, in the form first of Prussia, then of the Second German Reich, then of the Third Reich (the Nazi empire), played an important role in the region. There is no room here to consider the legacy of that complex experience. For some stimulating thoughts, see Mazower (2009b, esp. 553–604); Conrad (2012, 153–68).

countering Protestantism throughout their lands and beyond. In the Czech lands, they succeeded in suppressing Hussitism after the defeat of the Czech nobility at the Battle of Bíla Horá (White Mountain) in 1620.

Such a suppression was not possible in Hungary after the Ottoman occupation. The Ottomans in fact, as a counter to the Habsburgs, deliberately gave protection and support to Hungarian Protestants. Calvinism took particularly strong hold in southern and eastern Hungary, portions ruled by the Ottomans. The eastern city of Debrecen is still today called "the Calvinist Rome" and Protestantism is the most popular religion in that region of Hungary (Davies 1996, 494; Pálffy 2001, 120).

The Hungarian nobility also benefitted to a considerable extent from Ottoman rule. In 1556, Suleyman I established the autonomous Principality of Transylvania out of Hungary's eastern lands. This allowed the Hungarian Estates, following Habsburg accession to the diminished Kingdom of Hungary in 1526–1527, to maintain their independence from Vienna by threatening to transfer their allegiance to the rulers of Transylvania. "Due to their shrewd policy, in the seventeenth century the Hungarian Estates succeeded in preserving their privileges and the relative autonomy of the Hungarian Kingdom within the Habsburg Empire" (Pálffy 2001, 112). The Hungarian nobility suffered something of a loss of this freedom when, following the Ottoman expulsion from the Carpathian basin in the early eighteenth century, Transylvania was not returned to Hungary but was ruled directly from Vienna (until the Compromise of 1867).

With the re-imposition of Habsburg rule in the eighteenth century and the intermingling of the Hungarian and Austrian nobility, there was a natural strengthening of the Catholic element in Hungarian culture and a general tightening of Habsburg control. But the marks of imperial rivalry, and of the different imperial regimes, are still clear to see in Hungary today. Hungarians from the seventeenth century onwards began to recognize "two Hungaries", a "Western" and an "Eastern", Catholic and Protestant, pro-Habsburg (*labanc*) and independentist or pro-Transylvanian (*kuruc*). As late as the seventeenth century in the wars with the Habsburg-led coalition against the Ottomans, the largest Hungarian contingent fought on the side of the Ottomans. The Ottoman conquest "split the Hungarian soul in two", a split that continues to have repercussions today (Fodor 2013, 411).[12]

Thus, depending on what stance one takes today, the Ottomans—wrongly identified in the West as "Turks"—can be either a consoling or a menacing memory, though the balance seems to be swinging very much in the latter direction. For Pál Fodor the Ottoman conquest of Hungary was a virtually unmitigated disaster, a demographic and economic catastrophe that set Hungary back for centuries (Fodor 2013, 406–8). Géza Pállffy also calls Ottoman rule "an unmitigated tragedy", emphasizing in particular the depredations of warfare, "the fact that, for much of the century and a half, Hungary served as a battleground in the struggle between the forces of the great powers" (2001, 113–14, 124). Similarly for the current Hungarian leadership, with its strongly nationalist bent, the recent movement of refugees to Europe from the Middle East can be compared to the "Turkish" invasion of the sixteenth century and the refugees are referred to as "Turks" (many come via Turkey). Leaders warn of a new "Mohacs", the battle that decided the fate of Hungary.[13] Hungarians are used to being "Orientalized" by their western neighbors, but when it suits them they are very capable of using the same device towards perceived threats from "the East", and of portraying themselves, as they often did in the past, as bastions of European Christendom and of European values.

The Hungarians in this case reach backwards, across the recent Soviet period and even the period of interwar independence, to the experience of earlier empires. This is *longue durée* historical memory. Czechs today also

[12] The religious divisions and their imperial sources were vividly displayed in the great exhibition on the Protestant Reformation in Hungary at Budapest's National Museum in 2017 to mark the 500th anniversary of Martin Luther's *Theses.* Maps showed the areas of Ottoman rule and how they coincide with the current distribution of Protestants. It is made clear that it was Ottoman protection that made possible the survival of Protestantism in Hungary. Ottoman patronage of Protestantism was not merely political, though that was perhaps the main reason, as it wooed Protestant nations as allies against its principal enemies, the Catholic Habsburgs. But Ottoman thinkers also saw many parallels between Islam and Protestantism, for instance in the absence or diminution of church and clergy, and the hostility to making images. For the example of the close relation between the Ottomans and Protestant Elizabethan England, see Brotton (2016). John Elliott (1993) has also emphasized the extent to which Ottoman-Habsburg rivalry forced Catholic rulers, including the Habsburgs, to make concessions to local Protestant communities in order to get support for their wars. Ottoman pressure and presence therefore contributed to the making of European diversity.

[13] Kim Scheppele, intervention at the Conference of Europeanists, Philadelphia, April 2017.

look back to the distant past, in their case fondly, to the fourteenth-century reign of Charles IV, who was not just king of Bohemia but also Holy Roman Emperor, when the Czech lands can be held to have played a central role in European culture (Subrt 2018, 223–24).[14] But what is more striking is the relative absence today of references back to the earlier periods of imperial rule, the period of the Ottomans, the Habsburgs, the Romanovs. Such references were common during the period of Soviet control, when "Central Europe" and its characteristic values were discovered or re-discovered, often with reference back to the experience of those empires (Kumar 2001, 71–103). But since the fall of the Soviet empire, the overriding interest seems to be in tracing the legacy of that empire, and that empire alone, in the lives of the post-communist societies of East-Central Europe.[15]

One does still find references to the interwar period of independence, with both favorable and unfavorable connotations. Contemporary debates about refugees and immigrants provide one such point of comparison. For instance, Małgorzata Pakier and Joanna Wawrzyniak speak of the way in which "the prewar multiculturalism and the wartime fate of the Polish Jews has become a symbol to refer to when defending the 'others' of today" (Pakier and Wawrzyniak 2016a, 3). On the contrary, referring to

[14] Historical points of comparison—"golden ages", periods of greatness and grandeur, as well as of traumatic suffering and defeat—clearly vary greatly, depending on present circumstances. For instance, Czechs in the immediate aftermath of the failure of the Czech Spring of 1968 looked back nostalgically to the first Czech Republic of the Masaryk era; while in 1946 it was the Hussite Wars that were seen as the most significant (Subrt 2018, 226). For other examples of this variety, see Varga (2018).

[15] E.g., the essays in the volumes edited by Pakier and Wawrzyniak (2016b), Moskalewicz and Przybylski (2018), and—with some exceptions—Kopeček (2008). Pakier and Wawrzyniak, in proposing Eastern Europe as a "regional" framework of memory, argue that "making use of Eastern Europe as an umbrella concept makes sense since its societies were once influenced by a Soviet-type metanarrative and also by some resistance to it. Working through communism is thus very often a filter for other representations, especially for Fascism and Nazism" (2016a, 15). They are aware of course of earlier shared experiences in the region, but choose to focus only on the latest, the Soviet experience. This should be compared with earlier volumes on the region, e.g. Schöpflin and Woods (1989), Graubard (1991), where references back to the early empires are quite common. A recent volume edited by Omer Bartov and Eric Weitz (2013) also ranges within the earlier empires in searching for the roots of twentieth-century violence in the region.

It should be said that this narrowing refers mainly to academic "memory studies"; references to earlier imperial experiences—e.g. of Habsburg Galicia, for both Poles and Ukrainians—remain strong in popular and political discourse.

the hostility towards refugees shown by current Polish and Hungarian governments, George Soros accuses both Victor Orbán of Hungary and Jarosław Kaczynski of Poland of "trying to re-establish the kind of sham democracy that prevailed in the period between the First and Second World Wars in Admiral Horthy's Hungary and Marshall Piłsudski's Poland" (Soros and Schmitz 2016, 35–36; see also Hann 2015b, 103). Similarly, Jan Gross traces the negative attitudes of Poles towards immigrants today back to a longer tradition of Polish anti-Semitism, which he sees as expressing itself powerfully during the Second World War (in Hann 2015a, 2).

For Chris Hann, on the other hand, hostility towards refugees in Poland and Hungary can be attributed to the fact that "the former socialist countries had no exposure to the kind of capitalist multiculturalism which developed in most parts of Western Europe during the Cold War" (Hann 2015a, 2). So, we have explanations of anti-immigrant and anti-refugee attitudes in terms of pre-war experiences of anti-Semitism and populist authoritarianism on the one hand, and "really existing socialisms"—in the Soviet orbit—on the other. All this is to be expected: the past is a reservoir for comparisons and precedents of many kinds, often of a contradictory nature. It is also natural that it is the most recent past, where the residues and remains, material and moral, are most apparent, that is the primary focus of attention. Hence it is not surprising that it is the legacy of Soviet rule, as the latest imperial presence in the region, that comes in for the greatest scrutiny.

Shortly after the 1989 revolutions and the fall of the Soviet Union, Tony Judt stressed the continuities of East-Central European history:

> The Soviet-imposed regimes of Eastern Europe are part of their respective national histories; they continued in certain local traditions, pursued pre-established patterns of economic policy, and have contributed to the post-communist character of their societies. As with Pétain and Mussolini, so with the puppet authorities of the 'Peoples Democracies': however tempting it may be, they cannot be eliminated from their country's history, nor 'bracketed' from it, as an alien or passing aberration .... [T]he Communist experience did not come from nowhere, did not disappear without leaving a certain record, and cannot be written out of the local past, as it had earlier sought to extrude from that past elements prejudicial to its own projects." (Judt 1992, 102–3)

Judt did not wish to privilege the recent Communist experience, he merely sought to place it in the *longue durée* of East-Central European history. But even he does not point us to the centuries-long imperial experience, before the twentieth century, as providing possible clues to current concerns. The impact of the Soviet empire, and its continuing resonance in post-communist societies, has been examined in dozens of books and conferences. "Nostalgia" in the region, as a positive and negative affect, has been one of the liveliest and most interesting areas of such inquiry (e.g., Boym 2001; Todorova and Gille 2010; Hann 2012, 2015b; Angé and Berliner 2015). What though can one say about the earlier empires, and their possible continuing impact?

In his work, *Eastern Europe Between the Wars*, Hugh Seton-Watson observed that many of the states of the region had what he called "Western and Eastern" provinces, arising mainly out of former imperial experiences. "Special problems were created in provinces which had been ruled by 'Western' methods and, owing to the territorial settlement of 1918, came under 'Eastern' administration, or vice versa. Czechoslovakia, Yugoslavia, Romania, and Poland were states of mixed 'Western' and 'Eastern' provinces" (Seton-Watson 1967, 150). We have seen that Hungary recognizes "Western" and "Eastern" divisions, the result of rule under the Habsburgs and the Ottomans respectively. Similar divisions can be found in those countries that have experienced at various times, and in various regions, rule by Romanovs and Hohenzollerns as well as Habsburgs and Ottomans.

"Western" sections, it was held, were heir to relatively high levels of literacy and education, economic development and urbanization, "enlightened" administration and liberal institutions (Johnson 2002, 195–96). In cities like Prague and Budapest, Kraków and Lwów, Zagreb and Ljubljana—all in the Habsburg Empire—there was a flourishing public sphere and by the late nineteenth century the emergence of something like "civil society". Later, after 1918 and again in 1945, many of these cities and their regions found themselves in the "East", in states—Yugoslavia, Romania, the Soviet Union or its satellites—that had once been predominantly part of the Romanov or Ottoman Empires. There they joined cities and regions—Warsaw, Kyiv, Vilnius, Belgrade—that at least since the eighteenth century, had mostly missed out on the European Enlightenment and European Liberalism and where generally more repressive regimes held sway.

These distinctions of "West" and "East" are of course no more than "ideal types", probably too kind to the West and too censorious of the

East (there was, for instance, considerable "liberal" thought and culture in cities like Warsaw, Kyiv, and Vilnius, e.g., Jedlicki 1999). But like all ideal types there is enough truth in them to make them serviceable for many purposes. Let us take the example of Ukraine. In his well-known book *The Clash of Civilizations*, Samuel Huntington proclaimed that "the civilizational fault line between the West and Orthodoxy [seen by him as the symbol of Eastern Europe] runs through the heart [of Ukraine] and has done so for centuries" (Huntington 1997, 165). Like many others he distinguished between western Ukraine, centered on Lviv, and eastern Ukraine, centered on the Donetsk region and the Crimea, with Kyiv somewhere in between. The West has been predominantly Ukrainian speaking and strongly nationalist in outlook; it has also been mostly Uniate ("Greek Catholic"). The East has been mainly Russian speaking and Orthodox in religion (Plokhy 2015, 133–46).

The differences have shown themselves dramatically since the fall of communism: from the 1994 presidential elections, in which western Ukraine overwhelmingly supported the nationalist Leonid Kravchuk, and eastern Ukraine the pro-Russian Leonid Kuchma, to the more recent tragic events, in which Russia has annexed the Crimea and Ukraine is engaged in a long-running civil war, pitting the pro-European west against the pro-Russian east.

It is not too difficult to trace these differences to imperial legacies. Western Ukraine was for many centuries part of the Polish-Lithuanian Commonwealth then, from the time of the partitions of Poland, part of Habsburg Galicia. Lemberg/Lviv became the center of Ukrainian nationalism in the nineteenth century, partly, as we have seen, under the stimulus of the Austrian authorities. From the time of the Treaty of Riga (1921), it was again part of Poland, once more underlining its distance from the east. This made three million Ukrainian speakers citizens of the Second Polish Republic (Snyder 2003, 119–53; Plokhy 2015, 146–73, 215–27). After 1945, with Poland's enforced westward shift, it became part of Soviet Ukraine.

By contrast, eastern Ukraine ("left-bank" Ukraine) was from the time of Bohdan Khmelnytsky's rising against Polish rule (1654) part of the Russian Empire. After a brief period of independence (1917–1920) it was incorporated in the Soviet Union. Its Russianness has intensified over more than three centuries of imperial rule, just as Polish and Ukrainian culture have been predominant features of western Ukraine. Western Ukraine's leaning towards "Europe", the Europe of the European Union, was not necessarily part of its original conception of itself, although the Habsburg connection was no doubt important in this respect. But it has not proved difficult to make its Europhilia part of its identity and to sharpen the sense of its difference from the east. Poland, the historic enemy, with which it reached a rapprochement in the 1990s, has been replaced by Russia as the enemy (Snyder 2003, 252).

Poland itself has its "Western" and "Eastern" divisions, acknowledged in the well-known distinction between "Poland A" and "Poland B", with the Vistula as the line of demarcation. The divisions are historical, cultural, political and economic. Poland A, west of the Vistula, is more prosperous, more urbanized, and more economically developed. It includes the cities of Warsaw, Cracow, Wroclaw, and Poznan. Poland B, the less developed region east of the Vistula, includes the cities of Rzeszów, Lublin, Olsztyn and Białystok. The divisions have been reflected fairly closely in Polish politics since the fall of communism. Poland A has tended to support secular and liberal parties, such as the Democratic Left Alliance; Poland B has voted for socially conservative and more nationalist parties, such as the Law and Justice party.[16]

Again, as with Ukraine, the imperial background to this division is not hard to see. From the time of the eighteenth-century partitions, Poland A was shared by the Habsburg (Galicia) and the Hohenzollern (Silesia and Pomerania) Empires and benefitted from the greater freedom and economic progress in these empires. Poland B was for most of the time in the Russian Empire and suffered from the higher degree of repression and relative backwardness in this eastern region. Like the dye marks that show through repeated washings of clothes, the dye marks of the different empires have left enduring traces in the moral and material landscape of Poland. The Soviet empire, in this case, has by pushing Poland westwards accentuated the importance of Poland A, but also to some extent united the country in its common rejection of Soviet communism.

Perhaps we should think of the imperial experience in East-Central Europe as a palimpsest: a set of layers, one on top of each other, none of which completely obliterates what has gone before. Each empire—Ottoman, Habsburg, Hohenzollern, Romanov, Soviet—has contributed

[16] Some have also proposed adding a "Poland C", the plains and parks east of the Vistula, see *Gazeta Wyborcza* (1999). Again, it has to be emphasized that these distinctions and divisions are fairly crude, with many exceptions and variations. But their continuing popularity shows that they still have real significance.

its characteristic quality. Of course, the challenge is to identify the contribution of each. But the assumption has to be that even earlier periods of history have their persisting effects and that in human society nothing is ever completely lost, no memory completely obliterated.

"Europe" and Empire

There is one last way in which we might consider, however briefly, the imperial experience as a factor in East-Central Europe. Since the end of communism in the region, most of the countries have sought membership in the European Union and most have gained admittance. The "return to Europe" was one of the leading slogans of the 1989 revolutions. It animated the writings of many of the dissidents, as we saw with Milan Kundera's "The Tragedy of Central Europe" (Kumar 2001, 71–103). The experience of membership of the EU has been mixed for East-Central Europeans and led to a good deal of soul-searching as to national or regional identity. Not everyone is overjoyed at the "return to Europe", fearing the loss of certain distinctive East-Central European characteristics. But the alternative to membership seems equally unpalatable; the fate of Ukraine and Georgia in the face of a resurgent Russia seems to offer painful lessons. If the League of Nations was unable to play the buffer role originally fulfilled by the Habsburg Empire, perhaps the European Union—backed by NATO—might have more success.

But in opting for the European Union, or "Europe", is East-Central Europe once more accepting empire? Is the European Union simply empire in new clothes? That has been the argument of a number of commentators in recent years. Some have taken an optimistic view, regarding the European Union as a resurrection of the old Holy Roman Empire, an attractively loose confederation allowing for a high degree of autonomy and variation in the parts. In this view, East-Central Europeans have nothing to fear from membership of the EU. Their entry would do no more than add new elements of variety to what is already a variegated enterprise. Enlargement would mean enrichment to the mutual benefit of all members (Zielonka 2007).

There is a darker view, emanating mainly from the left but including also some on the nationalist right, that the EU is indeed a new empire but not of the benevolent type, like the Holy Roman Empire. For them the European Union is more akin to the overseas colonial empires of the Spanish, British, French, Dutch, Belgians and—for a while, in the late nineteenth century—the Germans. Eastern Europeans are being colonized, just as Asians and Africans were colonized in the nineteenth century, though perhaps less violently and with an apparent degree of consensus. They are, and probably will remain, second-class citizens, whatever their formal position. For commentators of this kind, the European Union is just a continuation of empire "by other means". Europeans did not give up on empire after 1945. They simply recreated it nearer home, as a species of "internal colonialism". East Europeans, in acceding to the EU, will simply exchange one form of imperial dependency for another. For them, enlargement means empire (see, e.g., Böröcz and Kovács 2001; Böröcz and Sarkar 2005).

There is no room here to discuss these provocative views. But they are just the latest indication of how much the shadow of empire hangs over East-Central Europe. It is as if true independence for the nations of this region is a fantasy, a myth, never to be realized. They were born, many of them, in empire, and there they must remain (similar views have been expressed about several sections of the post-colonial "Third World" or "Global South"). Their fitful bursts of freedom do not last. They are surrounded by more powerful neighbors who will always have designs on them, however innocent sounding their protestations. This is the true "tragedy of Central Europe".

This is of course an extreme, and in many ways severely distorted, view. No country or people are condemned to dependence. But by the same token there is no guarantee of freedom either. The peoples of East-Central Europe have for a large part of the last five hundred years lived under one empire or another. That in itself is not so unusual. Empire has been the experience of much of humanity for most of its history. Europe is no exception to that. In a global context, in world-historical time, there is nothing unusual about the experience of East-Central Europe.

But it was in Europe, from the eighteenth century onwards, if not earlier, that there was born the idea of national independence. Nations—however defined, which remains a contentious thing—were supposed to have their own states, to be ruled by their "own" people, their co-nationals. East-Central Europeans imbibed these ideas with even greater fervor than many of their neighbors, West and East (the English and the Russians for long remained skeptical about the national idea). It was here that they felt the greatest want, and the greatest failure when these wants were not met.

It is by comparison, not with the rest of the world, but with Spain, Britain, France, Germany, Italy, Russia and some others that the nations of

East-Central Europe feel that they have somehow missed out, that, unlike those, they have not been the masters of their own fates for much of their modern history. That can produce the characteristic sense of being victims, sometimes even of inferiority. It can also breed extremist responses, as it did particularly between the wars and as it seems to be doing in several places in the region currently. Wounded national pride can be a powerful stimulus to action, as Germany in particular showed in the 1920s and 1930s.

Empire has been a fact of life for East-Central Europeans for centuries. Its legacy can be clearly seen throughout the region, not least in the architecture of many of its major cities. Its peoples are struggling to come to terms with it, to make their own histories in their own way. What is still unclear is whether they will be allowed to, whether they will not be swallowed up in larger entities in which they will find themselves, once more, not the subjects but the objects of history.

References

Angé, Olivia, and David Berliner, eds. 2015. *Anthropology and Nostalgia*. New York and Oxford: Berghahn Books.

Bartov, Omer, and Eric D. Weitz, eds. 2013. *Shatterzone of Empires: Coexistence and Violence in the German, Habsburg, Russian, and Ottoman Borderlands*. Bloomington, IN: Indiana University Press.

Beller, Steven. 2006. *A Concise History of Austria*. Cambridge: Cambridge University Press.

Böröcz, Józef, and Melinda Kovács, eds. 2001. *Empire's New Clothes: Unveiling EU Enlargement*. Telfor, UK: Central European Review.

Böröcz, Józef, and Mahua Sarkar. 2005. What Is the EU? *International Sociology* 20, no. 2: 153–73.

Boym, Svetlana. 2001. *The Future of Nostalgia*. New York: Basic Books.

Brod, Max. 1947. *Franz Kafka, A Biography*, trans. G. Humphreys-Roberts. New York: Schocken Books.

Brotton, Jerry. 2016. *This Orient Isle: Elizabethan England and the Islamic World*. London: Penguin Books.

Conrad, Sebastian. 2012. *German Colonialism: A Short History*, trans. Sorcha O'Hagan. Cambridge: Cambridge University Press.

Crampton, R. J. 1997. *Eastern Europe in the Twentieth Century - And After*. 2nd ed. London: Routledge.

Davies, Norman. 1996. *Europe: A History*. Oxford: Oxford University Press.

———. 2011. *Vanished Kingdoms: The History of Half-Forgotten Europe*. London: Allen Lane.

Dénes, Ivan Zoltan ed. 2006. *Liberty and the Search for Identity: Liberal Nationalisms and the Legacy of Empires*. Budapest and New York: Central European University Press.

Elliott, J. H. 1993. Ottoman-Habsburg Rivalry: The European Perspective. In *Suleyman the Second and his Time*, ed. Halil Inalcik and Cemal Kafadar, 153–62. Istanbul: Isis Press.

Fodor, Pál. 2013. Hungary between East and West: The Ottoman Turkish Legacy. In *More Modoque. Die Wurzeln der europäischen Kultur und deren Rezeption im Orient und Okzident: Festschrift für Miklós Maróth zum siebzigsten Geburtstag*, ed. Pál Fodor, Gyula Mayer, Martina Monostori, Kornél Szovák, and László Takács, 399–419. Budapest: Argumentum Kiadó, MTA Bölcsészettudományi Kutatóközpont.

Gazeta Wyborcza. 1999. Polska A, B i C. August 4.

Gellner, Ernest. 1983. *Nations and Nationalism*. Oxford: Basil Blackwell.

Graubard, Stephen R. ed. 1991. *Eastern Europe ... Central Europe ... Europe*. Boulder: Westview Press.

Hann, Chris. 2012. Transition, Tradition, and Nostalgia: Postsocialist Transformations in a Comparative Framework. *Collegium Antropologicum* 36, no. 4: 1119–28.

———. 2015a. The Fragility of Europe's *Willkommenskultur*. *Anthropology Today* 31, no. 6: 1–2.

———. 2015b. Why Post-Imperial Trumps Post-Socialist: Crying Back the National Past in Hungary. In *Anthropology and Nostalgia*, ed. Olivia Angé and David Berliner, 96–122. New York: Berghahn.

Havel, Vaclav. 1987. The Power of the Powerless. In *Living in Truth*, ed. Jan Vladislav, 36–122. London: Faber and Faber.

Huntington, Samuel P. 1997. *The Clash of Civilizations and the Remaking of World Order*. New York: Touchstone Books.

Jedlicki, Jerzy. 1999. *A Suburb of Europe: Nineteenth-Century Polish Approaches to Western Civilization*. Budapest and New York: Central European University Press.

Johnson, Lonnie R. 2002. *Central Europe: Enemies, Neighbors, Friends*, 2nd ed. New York: Oxford University Press.

Judson, Pieter M. 2016. *The Habsburg Empire: A New History*. Cambridge: Harvard University Press.

Judt, Tony. 1992. The Past is Another Country: Myth and Memory in Postwar Europe. *Daedalus* 121, no. 4: 83–118.

Kann, Robert A. 1980. The Case of Austria. *Journal of Contemporary History* 15, no. 1: 37–52.

Kopeček, Michal. ed. 2008. *Past in the Making: Historical Revisionism in Central Europe after 1989*. Budapest: Central European University Press.

Kożuchowski, Adam. 2013. *The Afterlife of Austria-Hungary: The Image of the Habsburg Monarchy in Interwar Europe*. Pittsburgh: University of Pittsburgh Press.

Kumar, Krishan. 2001. *1989: Revolutionary Ideas and Ideals*. Minneapolis: University of Minnesota Press.

———. 2010. Nation-States as Empires, Empires as Nation-States: Two Principles, One Practice? *Theory and Society* 39 (2): 119–43.

———. 2017. *Visions of Empire: How Five Imperial Regimes Shaped the World*. Princeton: Princeton University Press.

Kundera, Milan. 1984a. The Tragedy of Central Europe. *New York Review of Books*, April 26: 33–38.

———. 1984b. An Interview with Milan Kundera. *Granta* 11: 21–37.

Magocsi, Paul Robert. 1995. *Historical Atlas of East Central Europe*. Seattle: University of Washington Press.

Manela, Erez. 2007. *The Wilsonian Moment: Self-Determination and the International Origins of Anticolonial Nationalism*. Oxford: Oxford University Press.

Mazower, Mark. 2009a. *No Enchanted Place: The End of Empire and the Ideological Origins of the United Nations*. Princeton: Princeton University Press.

———. 2009b. *Hitler's Empire: Nazi Rule in Occupied Europe*. London: Penguin Books.

Moskalewicz, Marcin, and Wojciech Przybylski, eds. 2018. *Understanding Central Europe*. Abingdon: Routledge.

Okey, Robin. 1992. Central Europe/Eastern Europe: Behind the Definitions. *Past and Present* 137: 102–33.

Pakier, Małgorzata, and Joanna Wawrzyniak. 2016a. Memory and Change in Eastern Europe: How Special? In *Memory and Change in Europe*, ed. Małgorzata Pakier and Joanna Wawrzyniak, 1–20. New York: Berghahn.

——— eds. 2016b. *Memory and Change in Europe: Eastern Perspectives*. New York and Oxford: Berghahn Books.

Pálffy, Géza. 2001. The Impact of the Ottoman Rule on Hungary. *Hungarian Studies Review* 28, no. 1–2: 109–32.

Pedersen, Susan. 2015. *The Guardians: The League of Nations and the Crisis of Empire*. Oxford: Oxford University Press.

Plokhy, Serhii. 2015. *The Gates of Europe: A History of Ukraine*. New York: Basic Books.

Roshwald, Aviel. 2001. *Ethnic Nationalism and the Fall of Empires: Central Europe, Russia and the Middle East, 1914–1923*. Abingdon: Routledge.

Rudling, Per Anders. 2015. *The Rise and Fall of Belarusian Nationalism, 1906–1931*. Pittsburgh: University of Pittsburgh Press.

Rupnik, Jacques. 1989. *The Other Europe*. New York: Pantheon Books.

Sayer, Derek. 2000. *The Coasts of Bohemia: A Czech History*. Princeton: Princeton University Press.

Schenk, Frithjof Benjamin. 2017. Eastern Europe. In *European Regions and Boundaries*, ed. Diana Mishkova and Balazs Trencsényi, 188–209. New York: Berghahn Books.

Schöpflin, George, and Nancy Woods, eds. 1989. *In Search of Central Europe*. Cambridge: Polity Press.

Seton-Watson, Hugh. 1967. *Eastern Europe Between the Wars, 1918–1941*. New York: Harper and Row.

———. 1980. Introduction to "Imperial Hangovers", Special Issue, *Journal of Contemporary History*, 15 (1): 1–4.

Snyder, Timothy. 2003. *The Reconstruction of Nations: Poland, Ukraine, Lithuania, Belarus, 1569–1999*. New Haven: Yale University Press.

———. 2010. *Bloodlands: Europe Between Hitler and Stalin*. New York: Basic Books.

Soros, George, and Gregor Peter Schmitz. 2016. 'The EU Is on the Verge of Collapse': An Interview. *New York Review of Books*, February 11: 35–37.

Stokes, Gale. 1997. *Three Eras of Political Change in Eastern Europe*. New York: Oxford University Press.

Subrt, Jiri. 2018. Historical Consciousness: Czech and Slovak Examples. In *Understanding Central Europe*, ed. Marcin Moskalewicz and Wojciech Przybylski, 222–28. London: Routledge.

Todorova, Maria, and Zsuzsa Gille, eds. 2010. *Post-Communist Nostalgia*. New York and Oxford: Berghahn Books.

Trenscényi, Balázs. 2017. Central Europe. In *European Regions and Boundaries*, ed. Diana Mishkova and Balazs Trencsényi, 166–87. New York: Berghahn Books.

Varga, Balint. 2018. My Hero, Your Enemy: Competing National Memory Cultures and Symbolic Politics in Central Europe. In *Understanding Central Europe*, ed. Diana Mishkova and Balazs Trencsényi, 156–62. New York: Berghahn Books.

Wolff, Larry. 2010. *The Idea of Galicia: History and Fantasy in Habsburg Political Culture*. Stanford: Stanford University Press.

Zielonka, Jan. 2007. *Europe as Empire: The Nature of the Enlarged European Union*. Oxford: Oxford University Press.

PART II

Political Regions of Memory

Partisan History and the East European Region of Memory

A. Dirk Moses

Introduction

Memory wars have been underway in Europe since the fall of the Soviet Union and recovery of national independence by East European countries, including the three Baltic states (Wawrzyniak and Pakier 2013). These wars are waged on various battle fronts: in domestic politics, interstate relations, and European Union agencies and forums. In the first, new national museums, memory institutes, and memorials depict these countries as successively invaded by the Soviets and the Nazis, victims of "double occupations" (Maier 2001–2002; Rohdewald 2008; Blaive et al. 2011). One of them—in Vilnius, Lithuania—even deploys a broad definition of genocide to advance the "double-genocide thesis," in which their countries were victims of two genocides, one by the Nazis, and another by the Soviets. In propagating such imagery, conservative political and cultural elites posit their nations as indigenous peoples occupied for much of

A. D. Moses (✉)
Department of Political Science, City College of New York
New York, NY, USA
e-mail: dmoses@ccny.cuny.edu

S. Lewis et al. (eds.), *Regions of Memory*, Palgrave Macmillan Memory Studies, https://doi.org/10.1007/978-3-030-93705-8_5

the twentieth century by their powerful neighbours—mainly by Germany and Russia—which attempted to subjugate and/or destroy them in various ways: by killing, deporting, and imprisoning designated political enemies and by importing Germans or Russian-speaking settlers.

The broad definition of genocide is inspired by theories of totalitarianism that circulated among anti-communist émigré activists. Not by coincidence, they echo the ideas of Raphael Lemkin (1900–1959), the anti-communist Jewish-Polish refugee scholar who coined the term in his book on the Nazi occupation of Europe during the Second World War, *Axis Rule in Occupied Europe* (1944). There he referred to the Nazi colonization of conquered territory in conceiving of genocide as a foreign occupation executed in a settler-colonial modality that attacked the cultural as well as the biological reproduction of a nation: genocide as a technique of occupation (Lemkin 1944).

The exclusion of cultural genocide from the United Nations Convention on the Punishment and Prevention of Genocide in 1948 has not stopped some of these new memory institutions utilizing the Lemkian definition of genocide. By linking genocide and occupation, they effectively mirror the "colonial turn" in the field of genocide studies, which drew on Lemkin to highlight the genocidal process affecting indigenous peoples in the classical settler colonies of Australia and North America (Moses 2008). These institutions likewise reflect the decolonizing trend in trauma studies by registering the debilitating impact of long-term but non-monumental repressions that usually attend foreign rule (Craps 2013).

To be sure, the death tolls and coercive demographic transformations in Eastern Europe make the invocation of genocide intuitively plausible. But this deployment is highly loaded in its connection with occupation because it posits the local Christian populations as indigenous people with authentic roots, despite the imperial and colonial traditions of some East European states, let alone the mythic status of such claims in a region of multiple migrations and hybridities. In this nativist reading, Jews count as settler outsiders despite their centuries-long presence and social integration. In the 1930s, nationalist Polish governments, for example, instrumentalized growing antisemitism to economically favour Christian Poles. Then, during the Soviet annexation of the eastern part of the country in 1939, they accused Jews of disloyalty due to their alleged Soviet sympathies, a pattern also discernible in the Baltic states. This framing fatally reproduces the logic that enabled the Holocaust in Eastern Europe, namely the collaboration between East European Christian nationalists and Nazi forces in eliminating the common "Judeo-Bolshevik" enemy (Zimmerman 2002; Himka and Michlic 2013; Hanebrink 2018). Rightwing East European diasporas and home governments similarly utilize the occupation and genocide concepts in a partisan manner against East European Jewish survivor minorities and Holocaust memory (Törnquist-Plewa and Yurchuk 2019). As we see in the section "Partisan Histories and the Partisan Subject" below, in doing so they invoke an ethic of cosmopolitan solidarity in European Union forums by claiming to unite Europeans as joint victims of occupation and genocide, whether by Nazis or Soviets ("double genocide"). Many commentators find this claim rings hollow when it entails displacing responsibility for the Holocaust entirely onto the Germans and blaming Jews for their victimization under the Nazis for supposedly supporting communism.

This instrumentalization of the genocide concept, which has roots in earlier émigré literature and academic work on communist crimes, has been roundly condemned by local and international critics, including from Israel (Budryte 2004; Courtois et al. 1999; Katz 2010; Freedland 2010; Zuroff 2010a). They point out the wilful blindness and/or calculated cynicism of claiming to be victims of genocide by the Soviets while occluding or underplaying the genocide their national heroes co-committed against Jewish fellow citizens immediately before and during the Nazi occupation. By contrast, at least since Jan Gross's pathbreaking book, *Neighbors: The Destruction of the Jewish Community in Jedwabne, Poland* (2000), writers, artists, politicians, scholars, curators, and activists in Poland and elsewhere in the region have been undertaking unflinching research on the Holocaust and its memorialization.

In response to these voices and their research, East European memory institutions have made adjustments to their exhibitions, and some countries have erected Holocaust memorials and Jewish museums. The Museum of the History of Polish Jews in Warsaw (POLIN) is a product of Jewish (local and diaspora) and state partnership, which evinces a sincere commitment to Jewish history in the longue durée history (https://www.polin.pl/en/). But they otherwise remain unimpressed by the insistence that remembrance and commemoration of Nazi crimes must trump Soviet ones. After all, East Europeans respond, their countries did not invade and attempt to destroy neighboring ones. Nor did their Russian, Western, and Israeli critics have to endure decades of communist rule, with its mass deportations, denial of national statehood for the Baltic states, and political and cultural repressions.

As always in this complex region of memory, local circumstances determine the memory coordinates. Because no significant Polish partisan movement proposed collaboration with the Nazis to confront Soviet rule, Polish nationalists commemorate victims of both rather than playing them off against one another. However, they all ask: where is the Western European recognition of this suffering? Nazis have been condemned and prosecuted for their crimes, and Holocaust memorials and museums exist to commemorate their victims. Communist genocides and mass crimes, complain many East Europeans, remains scandalously "unresolved" from a pan-European perspective, despite the demolition of communist monuments, renaming of streets, restitution, and lustration (Assmann 2011).

The debate has ground to halt with each side accusing the other of committing intellectual atrocities while trivializing actual ones. Mainly conservative East European politicians and intellectuals wield anti-totalitarian memory in the guise of the "double-genocide" thesis while facing, on one front, the Russian anti-fascist narrative that condemns their states as neo-fascist and, on the other, the established Western European conviction that the Holocaust is unique and incomparable (Kattago 2009).

In this chapter, I account for this European memory impasse via the notion of "partisan history" (Friedman and Kenney 2005). It has three related features, each gesturing to different semantic connotations of the word "partisan": first, partisan history refers to the East European nationalist partisans who fought the Soviet Union, itself the sponsor of the major wartime partisan forces that fought Axis forces; second, it represents partisan—that is, highly partial—arguments to protect the exalted status of these East European nationalist "freedom fighters"; and, third, its temporal structure collapses past and present so that contemporary nationalists imagine themselves to be partisans, weaponizing memory in fighting yesterday's battles today. The genocide concept is perfectly suited for partisan history. Lemkin did not consider that his creation could be put to ultra-nationalist rather than cosmopolitan uses, although genocide, as he conceived it, is an extreme policy of conquest and occupation of nations that inevitably provokes anti-colonial resistance. What is more, it is intrinsically racialist, because it presupposes the existence of contending races potentially engaged in racial warfare. As we see below, partisan history also has gendered consequences.

My argument has two further limbs. One, because East European memory (or memories, as they are always contested) extends to, and is influenced by, voices in Israel and North American diasporas, created by migration and refugee flows, the East European region of memory needs to include those places. Two, those partisan histories are not self-generated or autotelic, but products of tensions with overlapping regions of memory—with Russia, which since the fall of the Soviet Union has been attempting to fashion its own region of memory in Ukraine, Belarus, and other post-Soviet states (Kozachenko 2019; McGlynn 2020), and Western Europe. These interactions need to be included in the analytical frame. For instance, the East European imperative to adopt notions like double occupation and genocide are reactions to the hegemony of Holocaust memory in the pan-European memory field they entered after the Cold War, even leading to the adaptation of its tropes (Sierp 2014; Sierp and Wüstenberg 2015; Littoz-Monnet 2012; Stone 2012; Zombory 2017; Kovács 2018; Subotić 2019). The genocide concept is also particularly attractive for states that feel threatened by Russia and are emboldened by powerful ultra-nationalist diasporas (Finkel 2010, 57), while Russia deploys the concept in relation to ethnic Russian minorities in its "near abroad" (BBC 2021).

This reframing presents an alternative to the fantasy of a unified, continental, European region of memory that the European Union (EU) likes to advance (Mälksoo 2009). At the same time, it also transcends the customary East-West memory distinction by demonstrating how the European memory conflict is the product of overlapping regions of memory, and that the East European region of memory, with its preponderance of partisan histories, extends to Israel and North American diasporas. Once the partisan investments of all participants in overlapping, non-territorialized regions of memory are registered, I conclude, a non-partisan memory regime can be envisaged that transcends the zero-sum game in which the memory security for one side entails memory insecurity for the other (Mälksoo 2015). However, conditions within the region and its tense relations with neighbouring ones means that non-partisan history is difficult to realize in practice (Stone and Jinks 2022).

The chapter proceeds in four steps. First, it delineates the East European region of memory in relation to transnational processes of migration and diaspora-creation. For reasons of economy, it tends to generalize about the region and at the expense of its particularities. Second, it examines partisan histories and partisan political subjectivity. In part three, it addresses the partisan memory arguments and investments of the participants in the debate about mass crimes in twentieth-century Europe. The fourth section examines partisan criticisms of these histories by local, Western European, and Israeli critics.

East European Region of Memory

On the face of it, a stable and delimited East European region of memory seems plausible based on what the historian Charles S. Maier calls "territoriality," namely "the properties, including power, provided by the bordered political space, which until recently at least created the framework for national and often ethnic identity" (Maier 2000, 808). Whereas Western Europe was constituted by nation-states with relatively stable borders that suffered Nazi occupation only for a few years, Eastern Europe was constituted by unstable territoriality. "Small nations" predominated in what historians variously call a "shatterzone," "bloodlands," and "rimlands": the German and Russian empires in the north (Poland and the Baltic states), the Austro-Hungarian Empire with these neighbours in the middle (Poland, Ukraine, Czechia, Slovakia, Hungary), and to the south the Ottoman empire abutting the small Balkan states like Greece and Bulgaria (Bloxham 2008; Bartov and Weitz 2013; Snyder 2010; Levene 2013). This was the space that the English historian Lewis Namier in 1915 named "The European 'Middle East'" (Namier 1915, xiii). These borderlands of four Eurasian empires experienced recurrent cycles of invasion, civil war, ethnic cleansing, and genocide during the twentieth century: the violent substitution of multi-confessional empires by ambitious new states with contingent borders and populations, all too often based on mass expulsions of defenceless civilians, intent on domestic homogenization and in thrall to security paranoia. Traumatic collective experiences of forced migration and genocide produced partisan history. With this structural-historical difference came the attraction of rival "narratives of moral atrocity," one centred on the Holocaust in the West, another on imperialism and decolonization in Eastern Europe, much like the Global South. The former seeks to tame nationalism and Leviathan's potential for mass violence while the latter strives for territoriality to house post-Soviet nations (Maier 2000, 824).

And yet, regions of memory have transcended delimited geographical units wherever largescale movements of people constituted migration systems, namely enduring clusters of mobility—including reverse migration—that produced global diasporas connected to sending regions (Hoerder 2014; Assmann and Conrad 2010; Pakier and Wawrzyniak

2015). Such systems have constituted an element of proto-globalization for centuries, but dramatic European population growth, settler colonial expansion, and improvements in shipping and communications made the century between 1840 and 1940 an era of mass migration that greatly intensified the extent and scale of regions of memory. Diasporas extended regions of memory because migrants shuttled back and forth, exercising "flexible citizenship" and developing deterritorialized identities that encompassed an enlarged sense of home marked variously by hybridity and nostalgia, trauma, and loss (Agnew 2005).

East Europeans were no exception to this pattern. Between 1876 and 1910, poverty at home and labor demand in the Americas led to the emigration there of some 3.5 million members of the Habsburg monarchy alone: it was the leading supplier of migrants to the US (Zahra 2016). Many Ukrainian and Belarussian peasants from the Russian Empire—thus classed as Russians—went to Canada, Argentina, and Brazil (Kukushkin 2007). They—Christians and Jews—came in such numbers to the United States that nativists restricted this immigration in the Immigration Act of 1924, which also affected southern Europeans while excluding Asians entirely (Daniels 2004). In view of the squalor and exploitation they experienced in America's incipient industrial cities, about 30% to 40% of former Habsburg subjects returned (Zahra 2016, 14).

But most remained. Like other migrants, East Europeans brought with them memories of their diverse villages, towns, and cities to their new home—though not necessarily of a nation. Upon arrival, however, they were classified as members of a foreign region ("Eastern Europe," referring to former Habsburg and Romanov imperial territories) and nation ("Poles") awaiting assimilation (Stewart 1924). Over time, diasporic communities mirrored this national and regional consciousness: late-nineteenth-century Lithuanian national consciousnesses even developed more quickly among migrants to North American than in the homeland (Čiubrinskas 2010, 10). "Diasporic nationalism precedes homeland nationalism" as in some east Asian cases (Lei 2001, 360; Lei 2008, 182). For their part, post-imperial states of Czechoslovakia, Poland, Hungary, and Baltic countries embarked on national-building projects after the First World War, summoning their "nationals" abroad (though not Jews) to return (Zahra 2016, 114).

National and regional isomorphisms emerged from these coterminous and interacting local and transnational processes: the East European region of memory became inseparable from its diasporic imaginings, rooted not "in a primordial, bounded space" but in projections of homogeneous peoplehood that effaced emigrants' sub-national identities and languages. This "long-distance nationalism" (Glick-Schiller 2005) is captured by the "Lithuanian Charter" of the Supreme Committee for the Liberation of Lithuania, issued in 1949: its first clause includes the declaration that "Lithuanians scattered throughout the world make up one unified World Lithuanian Community" (LAC Charter 1949).

This notion of nationality was mirrored by the Zionist conception of Jews as a globally dispersed nation. East European Jews moved to many parts of the world, and their ideas about their birthplaces and their immigration destinations varied widely. Even so, Jewish migrants to North America from the Russian Empire did not join, say, Polish or Ukrainian organisations, but founded their own. Their relationship to Eastern Europe was also complicated by memories of poverty, antisemitic discrimination, pogroms, and then genocide. After the Second World War, and especially 1967, the affiliation to Eastern Europe as a Jewish homeland gradually moved to Israel (Kobrin 2010). The case of Israel is also a dramatic case of diaspora nationalism, created entirely by the Zionist diaspora, largely from Eastern Europe, even if also funded by western Jewish capital. Zionists agreed with East European nationalists that the so-called interwar "Jewish problem" could be solved to mutual advantage by mass emigration from Eastern Europe to Palestine (Zahra 2016, 18, 151).

It is no accident that seven of the first nine Israeli prime ministers hailed from the East European contact zone with Russia: David Ben-Gurion (born in Płońsk, Poland, then in the Russian Empire); Moshe Sharett (born in Kherson in today's Ukraine, then in the Russian Empire), Levi Eshkol (born in Orativ in today's Ukraine, then in the Russian Empire); Golda Meir (born Kiev in today's Ukraine, then in the Russian Empire); Menachem Begin (born in Brest in today's Belarus, then in the Russian Empire); Shimon Peres (born in Wiszniew, then Poland, in today's Belarus); and Yitzhak Shamir (born in Ruzhany in today's Belarus, then in the Russian Empire). By Zionism's East European origins and emigration from Russia and Ukraine of about 900,000 Jews since the 1990s, Palestine and then Israel became part of the East European region of memory (Zarembo 2017). Lapidary "diasporic intimacy" and intense "cultural intimacy" with Eastern Europe and Russia testify to diasporic Jewish and Israeli participation in the partisan histories that characterize this transnational region of memory (Boym 2001). Yad Vashem, the Israeli Holocaust memorial authority, for instance, initiated a major project called "Untold

Stories" that identifies and commemorates Holocaust massacre sites in the former Soviet Union, namely Belarus, Lithuania, Latvia, and Russia (Yad Vashem n.d.).

Christian diasporic communities in particular carried on the intense anticommunism of fledgling interwar nation-states after their forcible incorporation into the Soviet Union or Soviet sphere during and after the Second World War. The regional memory regime of the unstable and contested notion of "east-central Europe" thus became associated with decades of communist rule, overshadowing the comparatively few years of Nazi occupation. For them, communist rule stood for foreign (Russian) occupation and destruction of their national cultures that they had struggled for decades to emancipate from imperial entanglements and then house in states (Zake 2000). In these circumstances, diasporic nationalism enabled a purity-corruption binary in relation to homeland (Čiubrinskas 2010). Diasporic nationalists thus policed nation-ness by vetting aberrant identities, like immigrating Lithuanians "contaminated" by Soviet life or, earlier, Ukrainians from the Russian empire who did not necessarily identify as Ukrainian. Ukrainian diasporic organisations and their later historians in Canada ostracized them as socialist or Russophiles, limiting Ukrainian-ness to immigrants from Habsburg Galicia and Bukovyna (Čiubrinskas 2013; Kukushkin 2007, 7).

What is more, diaspora communities—particularly those in which emigrés and/or displaced persons (DPs) assumed a prominent role after the Second World War—came to regard themselves as depositories of authentic nation-ness while the homeland endured Soviet occupation, with its enforced secularism, Russification, and suppression of national independence. The "real" Lithuania lived on abroad during the siege of the homeland, cultivated in family life, churches, schools, festivals, and the discipline of language retention, ready for reimportation after liberation. Free to express the nationalist perspective in the West, diasporic community leaders formulated strategies to liberate the homeland from communist rule. Exiled Lithuanians and other Baltic leaders, for example, confronted the outright Soviet annexation of their countries by claiming that the mass deportations and repressions after the war constituted genocide—much as the framing of the deportations and massacres of Ottoman Armenians during the First World War as genocide was developed later by Armenians abroad, descendants of refugees and survivors (Budryte 2004). Likewise, the interpretation of the catastrophic famine in Ukraine in 1932–1933 as a Holodomor-genocide was developed by Ukrainian nationalists in North

America, and is promoted by descendants of the ultra-nationalist "insurgents" who sought an independent state against Soviet and Nazi forces, murdering tens of thousands of Poles and Jews in the process (Himka 2015). Anticommunist liberals in the US likewise advance this interpretation (Applebaum 2018).

In the main, East European partisan histories have been nurtured in the globalized East European region of memory. But what are partisan histories apart from ethnocentric memory patterns (Brown 2019)? Partisan history will feature in any regions of memory structured by competitive and uneven state formation, imperial competition, invasion, and mass violence. The Hungarian politician and scholar István Bibó, in an essay called "The Miseries of East European Small States," wrote about the region's unstable borders as "the major source of the political hysterias" (Bibó 2015, 137). He was approaching the question of "habitus" in a similar manner to the sociologist, Norbert Elias, who traced how German state formation influenced the development of national subjectivity and political culture (Elias 1996). Bibó goes so far as to write about how "The deformation of social structure was followed by the warping of the political self and a hysterical mental condition when there was no healthy balance between real, possible, and desirable" (Bibó 2015, 154). His observation is widely applicable. As the geographical space of the East European region of memory expands to Israel and diasporas, the imaginative space contracts and calcifies into partisan history. Fittingly, one scholar includes Israel in his category of existentially uncertain "small peoples" (Abulof 2009). As a region of memory of precarious small nations, it is characterized by "victimhood nationalism" (Lim 2010).

Partisan Histories and the Partisan Subject

The meaning of the highly mythologized term "partisans" depends on the context. From Spanish, Russian, and Polish perspectives, the partisan is an antifascist irregular who fought against Spanish Falangists, German and German-allied forces before and during the Second World War. The figure performs particular memory work in Russia and Yugoslavia where the masculinist image of the Slavic antifascist warrior overshadowed Jewish and female partisan contributions (Slepyan 2006; Pavasović Trošt 2018). And in the "Great Patriotic War," these irregulars joined ordinary soldiers and the massive Soviet civilian casualties as the commemorative focus, rather than the Holocaust, pairing immeasurable suffering with heroic resistance (Jones 2013). So pervasive was the suffering and resistance mythology that Soviet Belarus was known as a "partisan republic," a memory regime that one scholar has depicted as a "colonial discourse" imposed "on a dominated population," much of which had collaborated with the Germans during the war (Lewis 2017, 373–4; Goujon 2009). Communist partisans were also associated with national liberation in the Balkans; however, their ideal was Yugoslavian, whereas the Serbian nationalist Chetnik forces, which confronted both the German and the Soviet occupations, sought an independent nation-state. As might be expected, communist narratives about WWII downplayed, distorted, or suppressed the Chetnik narrative.

Since the end of the Cold War and disintegration of Yugoslavia into ethno-states, the anti-communist partisans have become the heroes for a different kind of partisan republic. The fall of communism between 1989 and 1991 is thus a turning point for partisan history: it is the moment when anticommunist East Europeans, not only rightwing nationalists, felt that they could finally tell their story. Whether pro- or anti-Soviet, heroization of partisans is a sign of partisan history. For the small nations and their diasporas in particular, memories of genocide and flight, and experiences of continuing exile from and occupation of imagined homelands, fuel "political hysterias," to invoke Bibó's term.

This experience is gendered in complex ways. On the one hand, it is tantamount to emasculation; the gendered coding of foreign occupation is indicated by phrases like "rape" of the nation and by the feminized status of ultimate victimhood, coveted for the sympathy and recognition of injustice it may elicit. On the other hand, the experience triggers its compensatory correction in a masculinist ethos of militarized resistance and national self-assertion to turn defeat into victory (Dibyesh 2007; Helms 2013; Barton Hronešová 2020). As a consequence, partisan history imagines politics as violent resistance: as "the act of taking the power to spill blood of the colonizer and using it himself," according to the Cameroonian theorist Achilles Mbembe (Mbembe 2017, 88). Its telos is redemption. In the case of the Holocaust, masculinist Zionist pioneers decried the supposed passivity of European Jews who went to their deaths "like sheep to the slaughter," resolving that they would "never again" be victims (Lim 2010, 147).

The resisting subjectivity of the colonized also becomes problematically racialized because it internalizes and turns back on the occupier the ideology of cultural difference and superiority that the latter used to justify their rule (Mbembe 2017, 88). The ensuing redemptive political projects of national homogeneity have grave consequences for ethnic and national minorities: ethnic cleansing and even genocide. Because Romanians blamed their nation's plight on colonization and cosmopolitanization by Jews, ridding Romania of Jews was a redemptive act of national liberation (Bejan 2006). Such redemptive projects are based on what Mbembe calls "a conspiratorial reading of history," namely a drama populated by the stock characters of "the executioner (enemy) and his victim (the innocent)": the former incarnate "the absolute form of cruelty," while the latter are "full of virtue … incapable of violence, terror or corruption" (Mbembe 2017, 88). There is only one plot in this drama: conniving enemies who seek to destroy the prostrate nation. The enemies are all the same, irrespective of identity, because partisan history knows only friends and enemies. Thus many Israelis regard Palestinians as Nazis, despite their manifestly different subject position and motives. The partisan expresses rather than transcends the binary logic of the colonial encounter (Mamdani 2004). He cannot see how history terrorizes him, and how he repeats the violence he endured in the name of ending foreign rule and domestic betrayal (Moses 2011).

The balance between suffering and resistance is played out in different ways depending on context. In Russia, consciousness of the suffering of domestic Stalinist victims is gradually being effaced by renewed heroic narratives about the Great Patriotic War: the victim Soviet Union prevailing over the perpetrator Nazis. The government's banning of the Memorial International NGO is only the most flagrant example of this trend (Etkind 2013; Kovács 2018; Roth 2021). In the Western European variant, the redemption inheres in inhabiting the two roles simultaneously: that of the perpetrator-collaborator with the Germans while also being their victim (Assmann 2011). In Eastern Europe, memory of partisans and deportation victims is complicated in different ways. In Serbia today, the antifascist partisans are accused of war crimes and their archenemies—the anticommunist Chetnik partisans—have been rehabilitated: Tito is now portrayed as a villain (Đureinović 2020). Reviled in the many decades of communist rule, anticommunist partisans regard themselves—and are regarded by many of their conationals—as sacrificial victims for the nation, their suffering embodying the collective trauma. Those who honor their martyrdom and continue their struggle in the present are similarly terrorized. "Fighting and suffering" are the common terms in Lithuania, for example, the former referring to the anti-Soviet partisans, while the latter includes deportation victims. Both embody the nation in part because of the postwar criminalization of partisans and silence about Soviet deportations and repression (Budryte 2004).

As embodiments of the nation in these post-communist states, partisans stand at the core of partisan history. The mission of the Genocide and Resistance Research Centre of Lithuania in Vilnius, for instance, is to memorialize the partisans who confronted the Soviets until 1953, as well as to study crimes committed on national soil:

> The study of genocide, crimes against humanity, and war crimes in Lithuania; the study of the persecution of local residents by occupying regimes; the study of armed and unarmed resistance to occupying regimes; the initiation of the legal evaluation of the activities of the organisers and implementers of genocide; and the commemoration of freedom fighters and genocide victims. (LGGRTC)

This formulation replaces an earlier hyperbolic one that authorized the Center to investigate "the physical and spiritual genocide of Lithuanians carried out by the occupying regimes between 1939 and 1990, and the resistance to the regimes," and also to "immortalise the memory of the freedom fighters and the genocide victims" (Lowe and Joel 2013, 78). This version reflected the mood that permeated the renewal of national consciousness in the late 1980s, namely during the *Perestroika* and *Glasnost* period, when space opened up to challenge the myth of Soviet liberation. Former partisans, deportees, and their families combined public remembrance of the deportations with the revival of calls for national liberation and independence that had been crushed by Soviet and Nazi occupations (Davoliūtė and Balkelis 2018). For many, the national self, recovering from the psychic shattering of that loss and suppression, imagined the new polity in ethnic terms as the titular nation's re-entry into history. The non-communist lineage required recovery, and the nation's suffering, hitherto a taboo under the Soviets, now demanded recognition. Rehabilitating the criminalized national partisans was central to this process.

In this context, the genocide concept was imported from the North American Lithuanian diaspora into the national struggle in Lithuania. Whereas exiles utilized the genocide concept to refer to deportations, local dissidents adopted it to refer to Russification. These dimensions were combined in the independence movement's adoption of the term. To appease public opinion, the communist government appointed a commission of inquiry into the deportations that used the term as well. After the fall of communism and regaining of national independence, the genocide thesis made its way into the Center for Research on Genocide and Resistance in 1994 (Budryte 2004).

The centrality of partisan memory to the national project continues. The official Lithuanian "Partisan Honouring, Military and Public Unity Day," for instance, collapses the anti-Soviet struggle of the 1940s and 1950s into the public campaign about the current Russian threat. The chief of the defence forces said that the day is to "honour those who fought for a free Lithuania and to show for the public what protectors and defenders of Lithuania's independence and freedom are today, how and with whom they are preparing for national defence and take part in military missions outside the country" (https://kam.lt/en/partisan_honouring_military_and_public_unity_day_celebration.html). For its part, Poland has instituted "National Remembrance Day of the 'Cursed Soldiers,'" which commemorates members of the underground Home Army who resisted the Sovietization of the country after war (Plocker 2022).

As deployed in official mythologies, partisan history performs discursive and emotional magic by casting spells of displacement to excuse violence at every turn. As a victimization fable, the colonized/partisan displaces responsibility to "external forces." Whatever his excesses, the partisan regards himself as the victim, always acting in national self-defence. If the partisan breaks the moral law, imposed circumstances gave him no choice. Persecution, deportation, and murder of minorities are justified as either understandable revenge (for, say, alleged Jewish collaboration with the Soviets) or, if they are later stigmatized, as incidental to the national project.

Significantly for post-communist states, partisan history addresses the narrative dilemma of soiled national foundations, namely the fact that partisans often collaborated with the Nazis in the hope of securing national independence, or that independent statehood was even a creature of Nazi design, as in Slovakia and Croatia. As detailed in the next section, partisan history disassociates national honor and the national project from Nazi contamination by rhetorical magic tricks like "deflective negationism" (Blutinger 2010). For example, Croatian authorities erected a Holocaust memorial in the national capital, Zagreb, to commemorate the "six million victims," on the spot that the Croatian-fascist deportation of local Jews took place, implying it was solely a German project (Vladisavljevic

2019). The same aim animated the abandoned "Holocaust law" in Poland, which proposed criminalizing references to "Polish death camps." Instead, the rightwing government wants to publicize the suffering of Poles in the "Polocaust" (Koposov 2017; Hackmann 2018; Gebert 2018). The former Ukrainian President Petro Poroshenko was pushing in the same direction when he announced at Yad Vashem in Israel that Ukraine was also a victim of genocide during the Second World War: "Ukraine, as a state that suffered from the Holodomor of 1932–1933, when millions of Ukrainians were tortured by the communist Stalinist regime that committed genocide against the Ukrainian people, reverently keeps the memory of the Holocaust victims as well" (Ahren 2019). And in late 2021, the Belarusian lower house passed a draft law, "On the Genocide of the Belarusian People," which seeks to criminalize the Nazi genocide of "the Belarusian people," conceived to include Jews and Christian citizens, thereby occluding the Holocaust in which Belarusians also participated (Rozovsky 2021).

Since 1991, East European partisan states control the future and the past, and (understandably) obsess about border security *vis-à-vis* Russia, the challenge on the eastern front. Security anxiety also extends to memory, which political scientists call "ontological security": "the idea that distinct understandings of the past should be fixed in public remembrance and consciousness in order to buttress an actor's stable sense of self as the basis of its political agency" (Mälksoo 2015, 222). The two forms of security are linked. It is no coincidence that the Ukrainian ultra-nationalists who fought against Soviet forces (and murdered Jews and Poles) are being honoured in Ukraine today, while the state is engaged in a struggle with Russian-backed separatists in its Donbass territory since 2014 (Dreyer 2018) and an invasion since February 2022. Yesterday's partisan struggles continue into the present. The battle over memory is existential for the partisan for another reason as well: because partisan history is so partial, the partisan subject is constitutively fragile. Inconvenient facts that expose partisan history's magic tricks are "normative threats" that need to be vanquished. The battle on the western front is Europe's prioritization of the Holocaust.

Double Genocide, Double Occupation, and the Uniqueness of the Holocaust

Many East European research centres, museums, and memorials embody and express the now well-known "double occupation" and "double genocide" theses, namely the proposition that the nations between Russia and Germany suffered two genocides, one against Jews by the Nazis and another against the majority population by the Soviets. An accentuated version of these theses codes Jews as the perpetrators of the initial Soviet genocide by casting them as supporters of the Soviet invasion and by depicting Bolshevism as essential Jewish. By emphasizing the crimes of communism, it is a species of totalitarianism theory that likewise affiliates Nazi and Communist regimes as similarly anti-liberal.

National Memory Politics

The three Baltic state museums adopt this conceptualization. The one in Vilnius undertakes study of the "Sovietisation of Lithuania," which began in 1940 when the "destruction of its political-social and economic structure, and cultural and traditional spiritual values ensued and enforcement of the Communist worldview and ideology in society started." The process continued after the Nazi occupation between 1941 and 1944 with "Soviet political and economic reprisals, terror, unjustified massacre of civilians (crimes against humanity), deportations, suppression of spiritual life, Russification of society, and coercive acts in modern Lithuanian history" (GRRCL). That is why the Vilnius center absorbed the "Museum of Occupations and Freedom Fights" in 1997 (known as the "Museum of Genocide Victims" until 2018), and why its Latvian and Estonian correlates are called the "Museum of the Occupation" (Weiss-Wendt 2008).

The language of empire saturates the exhibitions. Take the Latvian Museum of Occupation, which expresses anxiety about the Russian-speaking colonists who dilute the indigenous population after its wartime losses (Nollendorfs 2008, 211). It divides the century into three occupation experiences, each receiving equal attention; in the book of the museum of some 200 pages, the Holocaust receives two pages as an episode of the short Nazi occupation. To be sure, in the brief discussion of genocide, the murderous totality of the Holocaust is noted but Latvians also suffered genocide because the Soviet occupation aimed to substitute the indigenous culture with what it calls "A new breed of Russian-speaking, deracinated and culturally homogenized 'Soviet person'—*homo Sovieticus*" (Nollendorfs 2008, 108). Since 2018, a new law limits the use of Russian in public school instruction, much to the consternation of the Russian minority and neighboring Russia (Kim 2018).

In general, mourning is expressed for the consequent loss of cultural traditions and development opportunities. Consider the statement by Andrius Grikienis, a spokesman for Lithuania's mission to the EU, in 2010: "During the first years of Soviet occupation, Lithuania lost more than 780,000 of its residents. 444,000 fled Lithuania or were repatriated, 275,697 were deported to the gulag or exiled, 21,556 resistance fighters and their supporters were killed and 25,000 died on the front." In comparison, "More than 200,000 citizens of Jewish origin were killed by Nazis and their collaborators," he said. The point is clear: Lithuanians—a category that excludes Jews—suffered greater losses than Jews (Phillips 2010).

A description of the Memorial of the Victims of Communism and Anticommunist Resistance in Sighet, Romania, opened by a private initiative in 1997, communicates the basic sentiment.

> The Memorial Sighet was established as a reminder of the atrocities committed by the communist regime—for years the populace had been brainwashed to create the so-called "New Man" through the rewriting of history and poisoning the memories of generations. Moral and civic values could only be recovered if the collective consciousness is duly recuperated. Sighet prison was chosen because it was the first of many political prisons set up in Stalinist times and because it was where the country's political, spiritual and cultural elite of the pre-war democracy was exterminated. An International Study Centre was established here because out of all of the former communist countries, Romania's experience had been the longest and most painful—from the long years of suppressed resistance to Ceausescu's obscene "Golden Epoch". (Memorial Sighet 2013)

The museum is dedicated to the victims of totalitarianism, and the study centre to communism. Unlike most other East European states, Romania was not occupied by the Nazis, so its mass murder of Jews cannot be ascribed to the Germans. Accordingly, the memory fixation is on communism, and Jewish experiences are not really part of the equation.

Typical of partisan history here is the temporal framing: collapsing the Nazi and Soviet periods into one extensive category of undifferentiated "occupation regimes" that both committed genocide. In this operation, the Soviet "occupation" is implicitly more significant, because it attacked local Christian culture *for decades* whereas the short-lived Nazi presence murdered mainly those supposed Soviet allies, Jewish citizens. These meanings are contained less in memorials and monuments than in the discourse about them, which conveniently obscures the fact that Christian-national partisans—those immortal freedom fighters—were often the same forces that helped the Nazis exterminate local Jews or even took the lead, beginning the killing after the Soviets fled and before the Nazis arrived. This point is made by critics of the double genocide thesis in Lithuania in particular (Katz 2018). Similarly, when the Ukrainian Foreign Ministry invokes Lemkin as a supporter of its interpretation of the Holodomor-as-genocide, it refers to him as an "American lawyer" (neither Polish, nor Jewish) and refrains from linking Ukrainian partisan heroes to the Holocaust (MFA). Lemkin, who lost 49 members of his family in the Holocaust, would hardly have approved of this version of anti-colonial nationalism (Törnquist-Plewa and Yurchuk 2019; Lim 2021). Invoking genocide in this manner is partisan history.

The major crime of the twentieth century for partisan history, then, is the brutal occupation of the nation, first the Soviet occupation, interrupted by the brief Nazi one, followed by the lengthy Soviet domination until the early 1990s. For example, in the Vilnius genocide museum, a single room is dedicated to the Holocaust, and it is a relatively recent addition. For Poles, occupation was compounded by partition between the Soviets and Nazis in the Molotov-Ribbentrop pact, the perfidious epitome of totalitarian aggression against the smaller nations of the bloodlands. This experience, rather than the Holocaust, is the axis of the twentieth century. Since Jews are not classed as compatriots or co-nationals, their experiences hardly figure.

However scandalous this memory politics may appear to many in the West, it is explicable in light of the memory taboos that obtained under the Soviets. After all, one could not talk about the Holodomor until the late 1980s; nor has any Russian authority apologized or been prosecuted for mass crimes. Given the magnitude of the trauma, it is hardly surprising that in 2006 Ukrainian president Viktor Yushchenko established the correlate of the Israeli Yad Vashem: the Institute of National Memory, which assessed the Holodomor as genocide. Nor is it surprising that many in western Ukraine want to rehabilitate the reputation of wartime insurgents who fought for an independent Ukraine, a desire now vouchsafed by a law since 2015 ("On the Legal Status and Honoring the Memory of Fighters for Independence of Ukraine in the XX Century") that declares the Organisation of Ukrainian Nationalists (OUN) and Ukrainian Insurgent Army (UPA) as "independence fighters" and that criminalizes the questioning of their legitimacy (Katchanovski 2014; Luhn 2015). As a result, streets are renamed after their leaders who were antisemites and Nazi collaborators (Cohen 2017). The commemoration of the notorious Soviet execution of Ukrainian political prisoners in Lviv in June 1941 exhibits partisan memory by attributing the ensuing anti-Jewish pogrom to criminal elements rather than to Ukrainian nationalists (Himka 2015).

In Hungary, the Orbán government employs characteristically partisan gestures to deal with the uncomfortable fact that Hungary was a Nazi ally and that Hungarians enthusiastically worked with German occupiers to deport half a million Jewish Hungarians to Auschwitz in 1944 as part of their "Greater Hungary" war aims (Segal 2016). The 2011 constitution proclaimed that the country's independence was lost between March 1944 (the Nazi Occupation) and May 1990 (the end of communist rule), homogenizing the enemy and disavowing responsibility for the fate of Jewish Hungarians (Rév 2018, 610). Leading politicians and nationalist politicians narrated the fact that Hungarian soldiers fought the Soviets alongside Nazis as noble Christian crusade to defend the West against Bolshevism, a cause unconnected to the incidental wartime alliance. In fact, so this partisan history goes, Western Europeans abandoned the Western cause by siding with Stalin, leaving Hungary with little diplomatic room for manoeuvre. By virtue of the eventual German invasion, the country became the innocent victim of both anti-Christian, totalitarian regimes. What is more, there was no rupture in 1945, like in Western Europe: foreign occupation continued till the end of the communist regime. Ukrainian historiography of the war comes to similar conclusions (Dreyer 2018, 560). So does historiography in Croatia after 1992, though it is complicated by the criminalizing of the communist partisans and Serbian chetnik forces, heroization of the Croatian nationalists (that is, Ustasha fascists) who were confronted with "tragic choices" in their striving for national independence (Pavasović Trošt 2018).

The European Union

The same politics played out in European Union institutions. The "double genocide" thesis was popularized by East European political leaders and intellectuals in the "Prague Declaration on European Conscience and Communism" in 2008, and then given official recognition a year later in a European Parliament resolution on "European Conscience and Totalitarianism" that condemned "totalitarian crimes" (Neumayer 2019). The resolution is an outright challenge to the Western European belief in the uniqueness of the Holocaust, represented by an earlier declaration, the Declaration of the Stockholm International Forum on the Holocaust, signed by 46 governments in 2000. The Stockholm Declaration became the founding document of the International Holocaust Remembrance Alliance (IHRA), which is committed to promoting Holocaust memory as the foundation of a common European memory culture. As "unprecedented," declares the IHRA, the Holocaust "fundamentally challenged the foundations of civilization" and "will always hold universal meaning." Accordingly, "it must be forever seared in our collective memory" as a warning against "genocide, ethnic cleansing, racism, antisemitism and xenophobia" (IHRA). In this mode, Holocaust memory stands at the core of anti-racist education and genocide prevention. Passed 55 years after the liberation of the death camps, the Stockholm Declaration was regarded by EU elites as compensating for decades of official silence about the Holocaust. Already in 1995, they had debated making 27 January—the day in 1945 that Soviet troops entered Auschwitz—an official day in memory of the victims of the Holocaust, an idea taken up ten years later by the United Nations. In the place of failed efforts to provide an integrative EU memory based on common heritage, the Holocaust would now serve as Europe's foundation myth, the negation of its stated values (Diner 2004; Littoz-Monnet 2012).

This is not how East European states that emerged from communist rule saw matters. When ten of them joined the European Union in 2004, the opportunity arose to influence the continental memory regime. Conservative anticommunist governments among them could not regard 1945 and the liberation of the death camps as the European foundation moment because that year also marked the reimposition or commencement of nearly fifty years of Soviet domination. During 2008, conservative East European members of the European Parliament organised a conference and working group on "United Europe-United History" that led to the foundation of the "Reconciliation of European Histories Group," an informal all-party group of the European Parliament dominated by anticommunist East Europeans politicians. By reconciliation, it meant establishing a common approach to Nazi and Soviet crimes that sought equal treatment of all victims of totalitarian violence. As might be expected given its orientation, the group opposed a proposed EU ban on the Nazi swastika because not also banning Soviet symbols would represent a double standard. In the same year, the European Commission, then chaired by a conservative Slovenian government, held public hearings on "Crimes Committed by Totalitarian Regimes," and published a report coupling Nazi, fascist, and Stalinist crimes (Toth 2010, 8–10).

With supportive signals from the European Commission, which now sought to forge common European symbols, East European governments sponsored the 2009 European Parliament resolution on "European Conscience and Totalitarianism" that insisted on East European distinctiveness: those "countries have experienced both Communism and Nazism," meaning that "understanding has to be promoted in relation to the double legacy of dictatorship borne by these countries." Because the resolution had to gain a majority, it made passing reference to the "uniqueness of the Holocaust," but its burden was to counter the IHRA monopoly on the lessons of history by insisting that European integration and combatting "undemocratic, xenophobic, authoritarian and totalitarian ideas and tendencies" required the recognition of both Nazi and Communist crimes as Europe's "common legacy." It similarly undercut the centrality of antisemitism in the IHRA's anti-racism pedagogy by strategically adopting a generic victims' perspective to sideline perpetrator ideology in the evaluation of mass crimes: "from the perspective of the victims it is immaterial which regime deprived them of their liberty or tortured or murdered them *for whatever reason*" (EPD 2009, Emphasis added).

The non-hierarchical coupling of the Holocaust and Soviet crimes was continued in the "Platform of European Memory and Conscience," which was duly established as an NGO in 2011 pursuant to the EU Parliament resolution and the energy of the Polish Prime Minister, Donald Tusk, who then held the presidency of the European Parliament (Toth 2019, 17). Dominated by East European states, the platform is dedicated to "initiatives at the European level with a view to giving ***indiscriminate treatment*** to all crimes of genocide, crimes against humanity and war crimes, as well as to their victims" (PMEC. Emphasis added). The coupling of Nazi and Soviet crimes thus undercut the concession to the Holocaust's uniqueness in the 2009 resolution.

As an alternative pan-European day of remembrance, the Prague Resolution proposed the "European Day of Remembrance for Victims of

Totalitarian Regimes" on August 23, the anniversary of the Molotov-Ribbentrop Pact. According to the Polish Institute of National Remembrance, it aims to commemorate the victims of mass deportation and extermination, as well as to promote democracy, peace, and stability in Europe," and was first officially undertaken in Warsaw in 2011 "under the auspices of the Polish Presidency" (INR). Setting this date as a new European remembrance day was momentous: 2009 marked the seventieth anniversary of the notorious treaty between Nazi Germany and the Soviet Union that assigned the three Baltic states to the latter and divided Poland between them. The pact had long been remembered in the affected countries as the beginning of their subjugation, whether by Germans or Russians. Even those East European countries not subject by the pact, like the Czech Republic and Hungary, supported this date in the Prague Declaration, because it generally symbolized Nazi and Soviet hegemony that affected them all. What is more, the end of communism between 1989 and 1991 was referenced by its serial imposition as of 1939 (Sierp 2017).

Partisan Critiques of Partisan Histories

These partisan histories have attracted much criticism, especially from local Jewish communities and their non-Jewish supporters. A prominent critic is Vilnius-based Jewish Studies scholar Dovid Katz who collates these criticisms on his website, www.defendinghistory.com. Articles there excoriate the double genocide thesis and the barely concealed antisemitism that continues to link Jews to Bolshevism and now Russia. Why should Stalinist and Hitlerian crimes be remembered on the same day or in the same ceremony, he and the Israelis Ephraim Zuroff and Yehuda Bauer ask? These figures do not deny that Soviet crimes should be commemorated; but why conflate them with the Holocaust? In many publications, they expose the partisan histories of the region's successful national liberation movements that commemorate the Christian but not the Jewish dead and that wilfully cover up the fact that some of their national heroes were implicated in murdering Jews, and that their partisan heroic foundations of their national projects are thereby compromised. They also understandably denounce attempts to prosecute elderly former Soviet-Jewish partisans while not pursuing possible nationalist war criminals in their own ranks (Bauer 2010; Katz 2009, 2010, 2011; Zuroff 2010a,b).

From a Russian perspective, the Eastern Europe attack on the "Great Patriotic War" mythology is tantamount to blasphemy of a sacred memory that "Russians living abroad" (i.e., those who moved to East European countries under communist rule) should resist. Such resistance can be violent, as the riots attending the relocation of the Bronze Solider statue (originally "Monument to the Liberators of Tallinn") in Estonia in 2007 demonstrated (Wulf 2016). As might be expected, Russia's foreign minister attacked such moves and what he called the glorification of "the Nazis and their collaborators" (Russia Today 2012; Mälksoo 2013).

However telling these criticisms are, they too are manifestations of partisan history. For the Western European and Israeli claim about commemorating the Holocaust appropriately is not only that it is distinct in certain ways, but that it is metahistorically unique and represents a fundamental rupture of western civilization. Western civilization, so the reasoning goes, can only be reconstructed by placing memory of the Holocaust at the centre of its memory regime. Moreover, as we see with critics below, genocide should be defined in such a way that the Holocaust is effectively the only genocide in world history. This is also the position of Israel, a state founded by a national liberation movement, comprising a considerable number of Holocaust survivors, which also gained independent statehood with terroristic violence against alleged occupiers of the "homeland," in this case Palestinian Arabs and British forces. It is one thing for local Jewish communities to protest about the repression of the Holocaust and quite another for beneficiaries of ethnic cleansing with its own partisan memory regime.

A prominent proponent of this position is the historian Yehuda Bauer, born in Prague in 1926, and a Zionist partisan fighter (*Palmach*: elite "strike force" of the Jewish underground army) in the "war of national independence" in 1947 and 1948. He was an initiator of the Taskforce on International Co-operation on Holocaust Education, Remembrance and Research in the late 1990s that set its historical-philosophical agenda: "The Holocaust (Shoah) fundamentally challenged the foundations of civilization. The unprecedented character of the Holocaust will always hold universal meaning." The Holocaust, he continued, was a "conscious rebellion not just against the heritage of the Enlightenment, but against all the norms and traditions of Western civilization. Its utopia was a racist hierarchy, not any sort of egalitarianism" (Bauer 2009, 2010). This is a partisan proposition for any person who endured western colonialism with its claims about civilization and universal pretensions that justified racist utopias. It is only possible to make such statements about "civilization" by ignoring African and other anti-colonial thinkers who have long pointed to the violence of European colonialism as its condition of possibility that undermined Europe and the West's right to preach to others and to define what is uniquely evil.

These are not Bauer's views alone. Quoting his many writings on this subject, the Austrian historian Heidemarie Uhl declared in the *Israel Journal of Foreign Affairs* that "Indeed, Holocaust memory has become the historical foundation of the ethical and moral values in Western civilization, the basis of a Europe committed to human rights, to the struggle against racism, antisemitism, xenophobia and discrimination" (Uhl 2009). The tone is a new form of the civilizing mission that western Europeans have traditionally taken in relation to their apparently backward eastern neighbours (Wolff 1994; Mälksoo 2013, 182). The symptoms of the civilizing mission are the arrogation of the right to define civilization and to then to impose it on others in the name of universal rather than particular interests. Many western states are former colonial powers that deploy Holocaust memory in a partisan way, namely as a screen memory to cover over their own colonial crimes. Great Britain commemorates Holocaust Memorial Day, for example, while its government goes to great length to deny the violence it perpetrated in putting down national liberation movements in Asia and Africa in the 40s and 50s (Satia 2022).

The obfuscation of difficult knowledge about compromised pasts so typical of partisan history is also as apparent in Israel as in Eastern Europe. A law from 2011 enables the state to revoke funding for organisations that mourn Independence Day, as do Palestinian citizens of Israel, the memory NGO Zochrot, and cinemas screening films that incurred the ire of the culture and sports minister. The underground partisan forces that attacked Arabs are remembered in street names, as is the racist rabbi Meir Kahane who inspires the far-right in Israel (Azaryahu 1992; Ben-Ami 2011; Kolodney 2016). No partisan heroes from the war of independence were prosecuted for massacres and ethnic cleansing of Palestinians, a task that would be rendered most difficult by the systematic cleansing of Israeli archives. Documents that survived the coverup reveal looting—like cutting off fingers for rings—and raping usually associated with civilian and partisan attacks on Jews just a few years before (Shezaf 2019; Gross 2016). Menachem Begin and Yitzak Shamir were members of ultra-nationalist terrorist organisations (Begin of the Irgun and Shamir of the Irgun and then Lehi ["Stern Gang"]) in the 1930s and 1940s, and later became prime ministers of the country. In these respects, Israel exemplifies East European partisan history no less than Ukraine or Lithuania.

A further symptom of partisan ethnic memory is wilful blindness to Lemkin's broad definition of genocide, which was inspired by his Jewish religious upbringing, with its conviction that the nation was constituted as much by cultural memory and religious traditions greater than bare life (Moses 2021). Lemkin's invocation of "national spirit" also reflected the perspective of small nations, and is taken up by many East European states in their understanding of genocide. Observing that this legacy is being instrumentalized by Christian nationalist politicians and intellectuals, Dovid Katz and the Jewish-Lithuanian politician Leonidas Donskis restrict the definition of genocide to the Holocaust archetype. "Whether we like it or not," declared Donskis in 2009, "the Holocaust was the one and only bona fide genocide in human history … Ultimately, it was not a garden-variety mass killing." Donskis also sought to sever genocide from any imperial occupation nexus that Lemkin posited because, he continued, "we cannot regard the history of all our civilizations as one ongoing crime and one endless genocide of some group or other" (Donskis 2009). Such anthropological optimism is an expression of the philosophy that human civilization was progressing nicely until interrupted by the barbarism of the Holocaust.

Equally partisan is the trivialization of Soviet crimes. It is entirely reasonable for Bauer, Katz, and Zuroff to insist that the Soviet victory was a liberation for Jews. But it is partisan to suggest that it was necessarily a liberation for other members of this region because the Nazis were worse; that the Soviet crimes did not entail ethnic targeting, unlike the Holocaust, and that its repressive aspects were collateral damage or an unintended side effect; that the Soviet experience was somehow a progressive and idealistic if perverted project because locals could rise to positions of authority unlike Jews under the Nazis; and that the Soviets did not start the war when in fact they also invaded Poland (Zuroff 2010a; Mälksoo 2013, 188). In this manner, an English journalist railed against the double genocide thesis because, by any measure, the Holocaust surpassed Soviet crimes: "The oppression of the Soviet years was terrible, but it was not genocide: to be arrested is not to be shot into a pit. They are different and to say otherwise is to rob 'genocide,' a very specific term, of all meaning" (Freedland 2010). An Israeli journalist made the same point in insisting that "Stalin, with all its terrible crimes, did not develop a racial theory and did not engage in the systematic slaughter of peoples" (Primor 2012).

These are debatable assertions. Over eight million people—from many nationalities and social strata—died in Stalinist gulags and famines (Snyder 2011). That a racial theory makes the Nazi crimes more grave or significant is taken for granted rather than explained. Of course, there were differences between Nazism's apocalypticism—and ultimately, self-destructiveness—and later communism, which established a stable if brutal authoritarianism in the Brezhnev years, but that was after the mass killing of the Lenin and Stalin periods.

Insisting on Holocaust uniqueness is not a politically neutral or empirical assessment. It is a normative and highly partisan one that constitutes the standpoint for Israeli critics of East European memory regimes. Representative is the head of the Simon Wiesenthal Center's Israel office, Ephraim Zuroff, a religious-Zionist settler (migrating from the US diaspora in 1970) who lives in the illegal West Bank colony of Efrat. His presence there is possible only by military conquest, ethnic cleansing, and permanent occupation. His invocation of Holocaust uniqueness licences this positionality by claiming a special right of self-defence that blinds him to the invasive nature of his presence. "The history of the Middle East to date has shown that the best defence is a good offence," he said (Eesti Päevaleht 2002). It enables the same partisan ethno-nationalist politics as the East European he criticizes. The commitment to Greater Israel is no different to, say, the "Greater Hungary" or "Greater Romania" aspirations: militarily executed expansion in the name of recovering heritage and historical rights—naturally as an act of self-defence (Ibrahim 1990; Segal 2016; Solonari 2010).

The intersection of Holocaust uniqueness and expansion is nowhere more evident than in these extraordinary images of Israeli forces commemorating *Yom Hashoah* (Holocaust Memorial Day) as they demolished the residential tent in the Palestinian village of Susyia in the occupied West Bank before moving on to demolish edifices in the village of Um al-Khair (See Figs. 1 and 2).

Like East European partisans, these Israeli forces think they are driving out illegitimate populations—in this case, indigenous Palestinian villagers—who represent a supposed security threat or who stand in the way of realizing historical fantasies about homeland recovery and redemption. They engaged in extensive military violence to this end during the "war of independence" in 1947–1948 and now use legalized police coercion after conquest and occupation, as shown in these pictures. As noted above, for partisan forces, all opponents are the same: even hapless Palestinian

Fig. 1 Israeli military commemorate *Yom Hashoah* (Holocaust Memorial Day) as they demolish the residential tent in the Palestinian village of Susyia in the occupied West Bank. Image courtesy of Basil Adraa, May 2, 2019, with permission

villagers represent an existential threat and can be removed in good conscience inspired by Holocaust memory: never again will Jews be victims. Indeed, all Palestinians are demonized as mortal enemy, a "demographic timebomb," much as Baltic nationalists talk about the Russian-speaking minority. Zuroff criticizes the latter but not the former discourse because he is an exponent of partisan history.

Zuroff's declared intentions in denying the applicability of genocide to Soviet crimes also exemplifies partisan history. For example, he declares that Israel should not recognize the Holodomor as genocide because it lends credence to the claims of rightwing, antisemitic East Europeans (and Nazis) about Jews and communism. "If they [Soviet crimes] were [genocide], then that means that Jews committed genocide," he declared in a telling admission that attributes historical substance to the charge of "Judeo-Bolshevism." "There were Jews—not out of any loyalty to the Jewish people, and usually Jews who left the Jewish community—who worked in the KGB, in the Communist security apparatus, and did

Fig. 2 Israeli forces in the village of Um al-Khair. Image courtesy of Basil Adraa, May 2, 2019, with permission

horrible things. It's true" (Keinon 2019). To avoid the (in my view unnecessary) implication of Judeo-Bolshevism, he insists on placing the Holocaust at the apex of human suffering. The Israeli foreign ministry motive for likewise not recognizing the Holodomor as genocide has additional intensions, namely appeasing Russia (Sokol 2018).

Conclusion

These cases show also that just because the partisan memory has dirty hands does not mean its mutual criticisms are invalid. Such criticisms articulate the trauma of the victims of another national liberation movement: they correct each other's blind spots. Can a non-partisan history thus be imagined? Are East Europeans—including their diasporic communities and Israelis—able to face the criminality of their founding moments without flinching and playing partisan historical games? It seems possible when one considers that the emphasis on a titular population's sufferings is likely a compensation for their long repression in public remembrance under communism. In that sense, partisan historical theses like "double genocide" are forms of stigma management to negotiate ontological security in the face of normative remembrance hierarchies from WWII and its related genocidal practices (Adler-Nissen 2014; Mälksoo 2015). At the same time, Western Europeans would need to think in terms of learning processes rather than civilizing missions by imagining, say, the state of their nations had they been occupied by the Soviets since the 1940s with attendant persecutions and deportations. Non-partisan history and politics seem possible only when alternatives to anti-fascist and anti-totalitarian memory regimes evolve; when *all* victims—including colonial subjects—are mourned with the same emotional intensity that partisan liberation heroes enjoy (Gruber 2002; Gutman 2017).

Alas, conditions within the East European region of memory, and tensions between it and Western Europe and Russia, indicate that partisan history is unlikely to abate. A classic security dilemma, the memory security of one state represents an existential threat for another (Mälksoo 2015). Russia has been locked in a struggle with the Baltic states and Ukraine about the "Great Patriotic War" for years, and even invaded the latter in February 2022 to supposedly effect its "denazification." The Riga Museum of Occupation devotes its webpage to refuting Russian "fake news" about its partisan heroes, while Western Ukrainians rush to defend the reputation of Stepan Bandera, the leader of the Organization of Ukrainian Nationalists (OUN-B) militia that murdered Jews and Poles in an alliance of convenience with the German occupation of Soviet Ukraine during the Second World War. The Ukrainian government of the "orange revolution" had declared Bandera a "hero of Ukraine," and dedicated the year of 2019 to his honour (Rossoliński-Liebe 2014). In turn, Russia refers to the Ukrainian government as "Banderist" while refusing to acknowledge crimes the Soviet Union committed against East Europeans, let alone the Holodomor. In the Balkans, former belligerents engage in ranking victimhood—Jasenovac vs Vukovar vs Srebrenica—each standing as the "crime of crimes" for Serbs, Croats and Bosniaks respectively, each making claims to genocide and Holocaust terminology. Denial is part of this competition (Barton Hronešová 2021). Responding to the attempt of the outgoing international High Representative for Bosnia–Herzegovina Valentin Inzko to amend the state criminal code to criminalize denial of genocide, Bosnian Serb leaders engaged an Israeli Holocaust expert to cast doubt on Bosnian claims of genocide (Subotić 2022).

Partisan histories continue on the western front of the East European region of memory as well. A long-planned diplomatic initiative between

Israel and Visegrád states to strengthen rightwing populist forces in those countries was scuttled in early 2019 when Poland took umbrage at a statement by an Israeli politician about Polish antisemitism and culpability for the Holocaust (Heller 2019). Then on January 23, 2020, the Polish government boycotted the World Holocaust Forum meeting commemorating the 75th anniversary of Auschwitz's liberation for giving a platform to Russian president, Vladimir Putin, to criminalize neighbouring partisan republics as former Nazi collaborators while conveniently omitting mention of the Soviet role in partitioning Poland (Pfeffer 2020). Memory conflicts like these exemplify "political hysterias" observed by Bibó. The current stalemate repeats the partisan terms of discourse that led to mass violence in the first place.

Acknowledgement Thanks to Salvatore Babones, Jessie Barton Hronešová, Éva Kovács, Maria Mälksoo, Małgorzata Mazurek, Anat Plocker, Raz Segal, Dan Stone, Frances Tanzer, and the editors for helpful advice on drafts of this chapter. The usual disclaimers apply.

References

Abulof, Uriel. 2009. "Small Peoples": The Existential Uncertainty of Ethnonational Communities. *International Studies Quarterly* 53, no. 1: 22–48.

Adler-Nissen, Rebecca. 2014. Stigma Management in International Relations: Transgressive Identities, Norms, and Order in International Society. *International Organization* 68, no. 1: 143–176.

Agnew, Vijay. ed., 2005. *Diaspora, Memory and Identity: A Search for Home*. Toronto: University of Toronto Press.

Ahren, Raphael. 2019. Top Nazi Hunter Blasts Visiting Ukraine Leader for Ignoring Holocaust Complicity. *Times of Israel*, January 24.

Applebaum, Ann. 2018. *Red Famine: Stalin's War on Ukraine*. New York: Allen Lane.

Assmann, Aleida. 2011. *Auf dem Weg zu einer europäischen Gedächtniskultur*. Vienna. Picus Verlag.

Assmann, Aleida, and Sebastian Conrad, 2010. Introduction. In *Memory in a Global Age Discourses, Practices and Trajectories*, ed. Aleida Assmann and Sebastian Conrad, 1–17. Basingstoke: Palgrave Macmillan.

Azaryahu, Maoz. 1992. War Memorials and the Commemoration of the Israeli War of Independence, 1948–1956. *Studies in Zionism* 13, no. 1: 55–77.

Barton Hronešová, Jessie. 2020. *The Struggle for Redress Victim Capital in Bosnia and Herzegovina*. Basingstoke: Palgrave Macmillan.

Barton Hronešová, Jessie. 2021. Ethnopopulist Denial and Crime Relativisation in Bosnian Republika Srpska. *East European Politics* 38 (1): 21–42.

Bartov, Omer, and Eric D. Weitz, eds. 2013. *Shatterzone of Empires: Coexistence and Violence in the German, Habsburg, Russian, and Ottoman Borderlands.* Bloomington: Indiana University Press.

Bauer, Yehuda. 2009. Reviewing the Holocaust Anew in Multiple Contexts. *Jerusalem Center for Public History*, no. 80. May 1.

Bauer, Yehuda. 2010. Remembering Accurately on Int'l Holocaust Remembrance Day. *Jerusalem Post*, January 25.

BBC. 2021,"Russia-Ukraine: Putin compares Donbas war zone to genocide". *BBC News*, December 10. https://www.bbc.com/news/world-europe-59599066.

Bejan, Cristina. 2006. The Paradox of the Young Generation in Interwar Romania. *SLOVO, Interdisciplinary Journal of Russian, East-Central European and Eurasian Affairs*, 18, no. 2: 115–28.

Ben-Ami, Yuval. 2011. Myth and Murder in Israeli Street Names. June 26. https://972mag.com/myth-and-murder-in-israeli-street-names/17237/.

Bibó, István. 2015. The Miseries of East European Small States. in *The Art of Peacemaking: Political Essays by István Bibó*, ed. István Bibó and Iván Z. Dénes, 130–80. New Haven: Yale University Press.

Blaive, Muriel, Christian Gerbel, and Thomas Lindenberger, eds. 2011. *Clashes in European Memory: The Case of Communist Repression and the Holocaust.* Innsbruck, Studienverlag.

Bloxham, Donald. 2008. *Genocide, the World Wars and the Unweaving of Europe.* London: Vallentine Mitchell.

Blutinger, Jeffrey. 2010. An Inconvenient Past: Post-Communist Holocaust Memorialization. *Shofar: An Interdisciplinary Journal of Jewish Studies* 29, no. 1: 73–94.

Boym, Svetlana, 2001. *The Future of Nostalgia.* New York: Basic Books.

Brown, Kris. 2019. Political Commemoration and Peacebuilding in Ethno-National Settings: The Risk and Utility of Partisan Memory. *Peacebuilding* 7, no. 1: 51–70.

Budryte, Dovile. 2004. "We Call It Genocide": Soviet Deportations and Repression in the Memory of Lithuanians. In *The Genocidal Temptation: Auschwitz, Hiroshima, Rwanda, and Beyond*, ed. Robert S. Frey, 223–51. Lanham: University of America Press.

Čiubrinskas, Vytis. 2009, Diasporas Coming Home: Identity and Uncertainty of Transnational Returnees in Postcommunist Lithuania. In *Postsocialist Europe: Anthropological Perspectives from Home*, ed. László Kürti, Peter Skalník, 95–117. New York and Oxford: Berghahn Books.

Čiubrinskas, Vytis. 2010. Forging (Ethno)Nationalist Sentiment through Time and Space: Revivalist and Diasporic Ways of Accommodating Multiculturalism in East Europe and the USA. In *From Palermo to Penang / De Palerme a Penang: A Journey into Political Anthropology*, ed. Francois Ruegg and Andrea Boscoboinik, 103–8. Freiburg: Lit Verlag.

Craps, Steph, 2013. *Postcolonial Witnessing: Trauma Out of Bounds*. New York: Palgrave Macmillan.

Cohen, Josh. 2017. How Trump Can Show He's Tough on anti-Semitism. *Reuters*, June 20. https://www.reuters.com/article/us-cohen-antisemitism/commentary-how-trump-can-show-hes-tough-on-anti-semitism-idUSKBN19B2SS.

Courtois, Stéphane et al. 1999. *The Black Book of Communism: Crimes, Terror, Repression*. Cambridge, MA: Harvard University Press.

Daniels, Roger. 2004. *Guarding the Golden Door: American Immigration Policy and Immigrants Since 1882*. New York: Hill and Wang.

Davoliūtė, Violeta, and Tomas Balkelis, eds. 2018. *Narratives of Exile and Identity Soviet Deportation Memoirs from the Baltic States*. Budapest. CEU Press.

Dibyesh, Anand, 2007. Anxious Sexualities: Masculinity, Nationalism and Violence. *British Journal of Politics and International Relations* 9, no 2: 257–69.

Diner, Dan. 2004. Restitution and Memory: The Holocaust in European Political Cultures. *New German Critique* 90: 45–58.

Donskis, Leonidas. 2009. The Inflation of Genocide. *European Voice*, July 24. http://www.europeanvoice.com/article/2009/07/the-inflation-of-genocide/65613.aspx.

Đureinović, Jelena. 2020. *The Politics of Memory of the Second World War in Contemporary Serbia: Collaboration, Resistance and Retribution*. London: Routledge.

Dreyer, Nicolas. 2018. Genocide, Holodomor and Holocaust Discourse as Echo of Historical Injury and as Rhetorical Radicalization in the Russian-Ukrainian Conflict of 2013–18. *Journal of Genocide Research* 20, no. 4: 545–64.

Eesti Päevalehe. 2002. Dr. Efraim Zuroff online: inglisekeelsed vastused – Answers in English. *Eesti Päevalehe*, 8 August. https://epl.delfi.ee/online/b-dr-efraim-zuroff-online-inglisekeelsed-vastused-answers-in-english-b?id=50930740.

Elias, Norbert. 1996. *The Germans: Power Struggles and the Development of* Habitus *in the Nineteenth and Twentieth Centuries*. Cambridge: Polity Press.

European Parliament resolution of 2 April 2009 on European conscience and totalitarianism. http://www.europarl.europa.eu/sides/getDoc.do?pubRef=-//EP//NONSGML+TA+P6-TA-2009-0213+0+DOC+PDF+V0//EN.

Etkind, Alexander. 2013. *Warped Mourning Stories of the Undead in the Land of the Unburied*. Stanford: Stanford University Press.

Finkel, Evgeny. 2010. In Search of Lost Genocide: Historical Policy and International Politics in Post-1989 Eastern Europe. *Global Society* 24, no. 1: 51–70.

Freedland, Jonathan. 2010. I See Why "Double Genocide" is a Term Lithuanians Want. But it Appalls Me. *The Guardian*, 14 September.

Friedman, Max Paul, and Padraic Kenney, eds. 2005. *Partisan Histories: The Past in Contemporary Global Politics*. Basingstoke: Palgrave Macmillan.

Gebert, Konstanty. 2018. Projecting Poland and its Past. *Index on Censorship* 47, no. 1: 35–37.

Glick-Schiller, Nina. 2005. Long Distance Nationalism. In *Encyclopedia of Diasporas Immigrant and Refugee Cultures Around the World*, ed. Melvin Ember, Carol R. Ember, Ian Skoggard, 570–80. Boston, MA: Springer.

Goujon, Alexandra. 2009. Memorial Narratives of WWII Partisans and Genocide in Belarus. *East European Politics and Societies: and Cultures* 24, no. 1: 6–25.

Gross, Jan. 2000. *Neighbors: The Destruction of the Jewish Community in Jedwabne, Poland*. Princeton: Princeton University Press.

Gross, Jan. 2016. *Golden Harvest. Events at the Periphery of the Holocaust*. Oxford: Oxford University Press.

Gruber, Ruth Ellen. 2002. *Virtually Jewish: Reinventing Jewish Culture in Europe*. Berkeley: University of California Press.

Gutman, Yifat, 2017. *Memory Activism: Reimagining the Past for the Future in Israel-Palestine*. Nashville: Vanderbilt University Press.

Hackmann, Jörg. 2018. Defending the "Good Name" of the Polish Nation: Politics of History as a Battlefield in Poland, 2015–18. *Journal of Genocide Research* 20, no. 4: 587–606.

Hanebrink, Paul. 2018. *A Specter Haunting Europe The Myth of Judeo-Bolshevism*. Cambridge, MA: Harvard University Press.

Heller, Aaron, 2019. Israel hosts east European leaders after summit scrapped. AP News, February 19. https://www.apnews.com/a7d8eafb358b475aafa02691627c56fa.

Helms, Elissa. 2013. *Innocence and Victimhood: Gender, Nation, and Women's Activism in Postwar Bosnia-Herzegovina*. Madison: University of Wisconsin Press.

Himka, John-Paul. 2015. The Lontsky Street Prison Memorial Museum: An Example of Post-Communist Negationism. In *Perspectives on the Entangled History of Communism and Nazism: A Comnaz Analysis*, ed. Klas-Gőran Karlsson, Johan Stenfeldt and Ulf Zander, 137–66. Lanham: Lexington Books.

Himka, John-Paul, and Joanna Beata Michlic, eds. 2013. *Bringing the Dark Past to Light: The Reception of the Holocaust in Postcommunist Europe*. Lincoln: University of Nebraska Press.

Hoerder, Dirk. 2014. *Migrations and Belongings 1870–1945*. Cambridge, MA: Harvard University Press.

Ibrahim, Youssef M. 1990. In a West Bank Town, a Quiet Dialogue Between Settlers and Palestinians. *New York Times*, June 10.

Jambrek, Peter, ed. 2008. *Crimes Committed by Totalitarian Regimes.* Slovenian Presidency of the Council of the European Union. European Public Hearing on Crimes Committed by Totalitarian Regimes. Brussels.

Jones, Polly. 2013. *Myth, Memory, Trauma. Rethinking the Stalinist Past in the Soviet Union, 1953–70.* New Haven: Yale University Press.

Katchanovski, Ivan. 2014. The Politics of World War II in Contemporary Ukraine. *Journal of Slavic Military Studies* 27, no. 2: 210–33.

Kattago, Siobhan. 2009. Agreeing to Disagree on the Legacies of Recent History Memory, Pluralism and Europe after 1989. *European Journal of Social Theory* 12, no. 3: 375–95.

Katz, Dovid. 2009. On Three Definitions: Genocide; Holocaust Denial; Holocaust Obfuscation. In *A Litmus Test Case of Modernity. Examining Modern Sensibilities and the Public Domain in the Baltic States at the Turn of the Century*, ed. Leonidas Donskis, 259–77. Bern: Peter Lang.

Katz, Dovid. 2010. Why is the US Silent on "Double Genocide". *The Guardian*, 21 December.

Katz, Dovid. 2011. Detonation of the Holocaust in 1941: A Tale of Two Books. *East European Jewish Affairs* 41, no, 3: 207–21.

Katz, Dovid. 2018. The Baltic Movement to Obfuscate the Holocaust. In *Mass Violence in Nazi-Occupied Europe*, ed. Alex J. Kay and David Stahel, 235–61. Bloomington: Indiana University Press.

Keinon, Herb. 2019. Zuroff: Israel Should Not Recognize Holodomor as Genocide. *Jerusalem Post.* January 22.

Kim, Lucian, 2018. A New Law in Latvia Aims to Preserve National Language by Limiting Russian in Schools. *NPR*, October 28. https://www.npr.org/2018/10/28/654142363/a-new-law-in-latvia-aims-to-preserve-national-language-by-limiting-russian-in-sc.

Kobrin, Rebecca. 2010. *Jewish Bialystok and its Diaspora.* Bloomington: Indian University Press.

Kolodney, Ziva. 2016. Contested Urban Memoryscape Strategies and Tactics in Post-1948 Haifa. In *The War of 1948: Representations of Israeli and Palestinian Memories and Narratives*, ed. Avraham Sela and Alon Kadish, 101–20. Bloomington: Indiana University Press.

Koposov, Nikolay. 2017. *Memory Laws, Memory Wars: The Politics of the Past in Europe and Russia.* Cambridge. Cambridge University Press.

Kovács, Éva. 2018. Limits of Universalization: The European Memory Sites of Genocide. *Journal of Genocide Research* 20, no. 4: 490–509.

Kozachenko, Ivan. 2019. Fighting for the Soviet Union 2.0: Digital Nostalgia and National Belonging in the Context of the Ukrainian Crisis. *Communist and Post-Communist Studies* 52, no. 1: 1–10.

Kukushkin, Vadim. 2007. *From Peasants to Labourers: Ukrainian and Belarusan Immigration from the Russian Empire to Canada.* McGill-Queens University Press.

LAC Charter. 1949. http://lithuanian-american.org/home/structure/rules-regulations/.

Lemkin, Raphael, 1944. *Axis Rule in Occupied Europe: Laws of Occupation, Analysis of Government, Proposals for Redress.* Washington, DC: Carnegie Endowment for International Peace.

Levene, Mark. 2013. *The Crisis of Genocide: Devastation*, Vol. 1. The European *Rimlands* 1912–1938. Oxford: Oxford University Press.

Lewis, Simon, 2017. The "Partisan Republic": Colonial Myths and Memory Wars in Belarus. In *War and Memory in Russia, Ukraine, and Belarus*, ed. Julie Fedor et al., 371–96. Houndmills: Palgrave Macmillan.

Lie, John. 2001. Diasporic Nationalism. *Cultural Studies<-->Critical Methodologies* 1, no. 3: 355–62.

Lie, John. 2008. *Zainichi (Koreans in Japan): Diasporic Nationalism and Postcolonial Identity.* Berkeley and Los Angeles: University of California Press.

Lim, Jie-Hyu. 2010. Victimhood Nationalism in Contested Memories: National Mourning and Global Accountability. In *Memory in a Global Age Discourses, Practices and Trajectories*, ed. Aleida Assmann and Sebastian Conrad, 138–62. Basingstoke: Palgrave Macmillan.

Lim, Jie-Hyu. 2021. Triple Victimhood: On the Mnemonic Confluence of the Holocaust, Stalinist Crime, and Colonial Genocide. *Journal of Genocide Research* 23, no. 1: 105–26.

Littoz-Monnet, Annabelle. 2012. The EU Politics of Remembrance: Can Europeans Remember Together? *West European Politics* 35, no. 5: 1182–202.

Lowe, David, and Tony Joel. 2013. *Remembering the Cold War Global Contest and National Stories.* London: Routledge.

Luhn, Alex. 2015. Ukraine Bans Soviet Symbols and Criminalises Sympathy for Communism. *The Guardian*, May 22, 2015.

McGlynn, Jade. 2020. Historical framing of the Ukraine Crisis through the Great Patriotic War: Performativity, Cultural Consciousness and Shared Remembering. *Memory Studies* 13, no. 6:1058–80.

Maier, Charles S. 2000. Consigning the Twentieth Century to History: Alternative Narratives for the Modern Era. *American Historical Review* 105, no. 3: 807–31.

Maier, Charles S. 2001–2002. Heißes und Kaltes Gedächtnis: Über die politische Halbwertszeit von Nazismus und Kommunismus. *Transit* 22: 153–65.

Mälksoo, Maria. 2009. The Memory Politics of Becoming European: The East European Subalterns and the Collective Memory of Europe. *European Journal of International Relations* 15, no. 4: 653–80.

Mälksoo, Maria. 2013, "Nesting Orientalisms at War: World War II and the 'Memory War' in Eastern Europe." In *Orientalism and War*, ed. Tarak Barkawi, and Keith Stanski, 177–95. New York: Columbia University Press and Hurst.

Mälksoo, Maria. 2015. "Memory Must be Defended": Beyond the Politics of Mnemonical Security. *Security Dialogue* 46, no. 3: 221–37.

Mamdani, Mahmood. 2004. *When Victims Become Killers: Colonialism, Nativism, and the Genocide in Rwanda*. Princeton: Princeton University Press.

Mbembe, Achille. 2017. *Critique of Black Reason*. Durham: Duke University Press.

Memorial Sighet. 2013. Prison of the Ministers (Memorial Sighet). http://www.beyondtheforest.com/Romania/RSR6.html.

Moses, A. Dirk, ed. 2008. *Empire, Colony, Genocide: Conquest, Occupation, and Subaltern Resistance in World History*. New York and Oxford: Berghahn Books.

Moses, A. Dirk. 2011. Genocide and the Terror of History. *Parallax* 17, no. 4: 90–108.

Moses, A. Dirk. 2021. *The Problems of Genocide: Permanent Security and the Language of Transgression*. Cambridge: Cambridge University Press.

Namier, Lewis, 1915. *Germany and Eastern Europe*. London: Duckworth.

Neumayer, Laure. 2019. *The Criminalisation of Communism in the European Political Space after the Cold War*. Abingdon and New York: Routledge.

Nollendorfs, Valters. 2008. *Museum of the Occupation of Latvia, 1940–1991*. Riga: Museum of the Occupation of Latvia.

Pakier, Malgorzata, and Joanna Wawrzyniak. eds. 2015. *Memory and Change in Europe: Eastern Perspectives*. New York and Oxford: Berghahn Books.

Pavasovic Trost, Tamara. 2018. Ruptures and Continuities in Nationhood Narratives: Reconstructing the Nation through History Textbooks in Serbia and Croatia. *Nations and Nationalism* 24, no. 3: 716–40.

Pfeffer, Anshel, 2020. In new battle over Auschwitz legacy, Poland falls victim to Holocaust geopolitics. *Ha'aretz*, 22 January.

Phillips, Leigh. 2010. EU Rejects Eastern States' Call to Outlaw Denial of Crimes by Communist Regimes. *The Guardian*, December 21.

Plocker, Anat. 2022. *The Expulsion of Jews from Communist Poland: Memory Wars and Homeland Anxieties*. Bloomington: Indiana University Press.

Primor, Ada. 2012. Hitler's Little Mustache. *Ha'aretz*. February 14, 2012.

Rév, István. 2018. Liberty Square, Budapest: How Hungary Won the Second World War. *Journal of Genocide Research* 20, no. 4: 607–23.

Rohdewald, Stefan. 2008. "Post-Soviet Remembrance of the Holocaust and National Memories of the Second World War in Russia, Ukraine and Lithuania." *Forum for Modern Language Studies* 44, no. 2: 173–84.

Rossoliński-Liebe, Grzegorz. 2014. *Stepan Bandera: The Life and Afterlife of a Ukrainian Nationalist. Fascism, Genocide, and Cult*. Stuttgart: ibidem.

Roth, Adrian. 2021. Russia moves to shut down country's most celebrated civil rights group. *The Guardian*, February 12. https://www.theguardian.com/world/2021/nov/11/russia-moves-to-shut-down-countrys-most-celebrated-civil-rights-group.

Rozovsky, Liza. 2021. Belarus under fire for law equating Holocaust with Nazi crimes against nationals. *Ha'aretz*, December 19. https://www.haaretz.com/misc/article-print-page/.premium-israeli-historians-belarus-blurring-holocaust-with-genocide-legislation-1.10477485.

Russia Today. 2012. Moscow Enlists Russian Expats in Fight Against Historical Revisionism. October 26. http://rt.com/politics/russia-lavrov-history-war-expatriates-282.

Satia, Priya. 2022. Britain's Culture War: Disguising Imperial Politics as Historical Debate about Empire. *Journal of Genocide Research* 24, no. 2: 308–20.

Segal, Raz. 2016. *Genocide in the Carpathians: War, Social Breakdown, and Mass Violence, 1914–1945*. Stanford: Stanford University Press.

Shezaf, Hagar. 2019. Burying the Nakba: How Israel Systematically Hides Evidence of 1948 Expulsion of Arabs. *Ha'aretz*, July 4, 2019.

Sierp, Aline. 2014. *History, Memory and Trans-European Identity: Unifying Divisions*. New York. Routledge.

Sierp, Aline. 2017. 1939 versus 1989—A Missed Opportunity to Create a European *Lieu de Mémoire*? *East European Politics and Societies and Cultures* 31, no. 3: 439–55.

Sierp, Aline, and Jennifer Wüstenberg. 2015. Linking the Local and Transnational: Rethinking Memory Politics in Europe. *Journal of Contemporary European Studies* 23, no. 3: 321–29.

Slepyan, Kenneth. 2006. *Stalin's Guerrillas: Soviet Partisans in World War II*. Lawrence, KS: University of Kansas Press.

Snyder, Timothy, 2011. Hitler vs. Stalin: Who Killed More? *New York Review of Books*, March 10.

Sokol, Sam. 2018. The Tension between Historical Memory and Realpolitik in Israel's Foreign Policy. *Israel Journal of Foreign Affairs* 12, no. 3: 311–24.

Solonari, Vladimir. 2010. *Purifying the Nation, Population Exchange and Ethnic Cleansing in Nazi-allied Romani*. Washington, DC and Baltimore: Woodrow Wilson Center Press and the Johns Hopkins University Press.

Stewart, Ethelbert. 1924. The New Immigration Quotas, Former Quotas, and Immigration Intakes. *Monthly Labor Review* 19, no. 2: 1–11.

Stone, Dan. 2012. Memory Wars in the New Europe. In *The Oxford Handbook of Postwar European History*, ed. Dan Stone. 714–31. Oxford: Oxford University Press.

Stone, Dan, and Becky Jinks, 2022. Genocide and Memory. In *Genocide: Key Themes*, ed. Donald Bloxham and A. Dirk Moses. Oxford: Oxford University Press.

Subotić, Jelena. 2019. *Yellow Star, Red Star: Holocaust Remembrance after Communism*. Ithaca: Cornell University Press.

Subotić, Jelena. 2022. Holocaust and the Meaning of the Srebrenica Genocide: A Reflection on a Controversy. *Journal of Genocide Research* 24, no. 1: 71–82.

Snyder, Timothy. 2010. *Bloodlands: Europe Between Hitler and Stalin*. New York: Basic Books.

Törnquist-Plewa, Barbara, and Yuliya Yurchuk. 2019. Memory Politics in Contemporary Ukraine: Reflections from the Postcolonial Perspective. *Memory Studies* 12 no. 6: 699–720.

Toth, Mano. 2019. Challenging the Notion of the East-West Memory Divide. *Journal of Common Market Studies* 57, no. 5:1031–50.
Uhl, Heidemarie. 2009. Conflicting Cultures of Memory in Europe: New Borders Between East and West. *Israel Journal of Foreign Affairs* 3, no. 5: 60–72.
Vladisavljevic, Anja. 2019. Zagreb Prepares New Monument to Holocaust Victims. *Balkan Insight*, June 4. https://balkaninsight.com/2019/06/04/zagreb-prepares-new-monument-to-holocaust-victims.
Wawrzyniak, Joanna, and Malgorzata Pakier. 2013. Memory Studies in Eastern Europe: Key Issues and Future Perspectives. *Polish Sociological Review* no. 183: 257–79.
Weiss-Wendt, Anton. 2008. Victim of History: Perceptions of the Holocaust in Estonia. *Journal of Baltic Studies* 39, no. 4: 89–104.
Wolff, Larry. 1994. *Inventing Eastern Europe: The Map of Civilization on the Mind of the Enlightenment*. Stanford: Stanford University Press.
Wulf, Meike. 2016. *Shadowlands. Memory and History in Post-Soviet Estonia*. New York and Oxford: Berghahn Books.
Yad Vashem. n.d. "The Untold Stories: The Murder Sites of Jews in the Occupied Territories of the Former Soviet Union." https://www.yadvashem.org/untoldstories/database/homepage.asp.
Zahra, Tara. 2016. *Great Departure: Mass Migration from Eastern Europe and the Making of the Free World*. New York: W.W. Norton.
Zake, Ieva. ed. 2000. *Anti-Communist Minorities in the U.S.: Political Activism of Ethnic Refugees*. Basingstoke: Palgrave Macmillan.
Zarembo, Kateryna. 2017. Diplomatic Tightrope and Historical Memory: Ukraine-Israel Foreign Policy Audit. *Euromaiden Press*, June 22. http://euromaidanpress.com/2017/06/22/diplomatic-tightrope-and-historical-memory-ukraine-israel-foreign-policy-audit/.
Zimmerman, Joshua, ed. 2002. *Contested Memories: Poles and Jews during the Holocaust and its Aftermath*. Piscataway, NJ: Rutgers University Press.
Zombory, Máté. 2017. The Birth of the Memory of Communism: Memorial Museums in Europe. *Nationalities Papers* 45, no. 6: 1028–46.
Zuroff, Ephaim. 2010a. A Dangerous Nazi-Soviet Equivalence. *The Guardian*, 29 September.
Zuroff, Ephaim. 2010b. US Must Take a Tougher Line on Baltic Revisionism. *The Guardian*, 29 November.

China, the Maritime Silk Road, and the Memory of Colonialism in the Asia Region

Laura Pozzi

At the beginning of the twenty-first century, the People's Republic of China (PRC) emerged as one of the major political powers in the Asia region, and as such its influence over the social memory of Asia's past is gradually increasing. Projects such as the "One Belt, One Road" (*yidai yilu* 一带一路), launched in September 2013 by president Xi Jinping 習近平, exemplify China's attempt to strengthen its political and financial influence in Asia, also through the shaping of transnational historical memories of the Asia region.

This chapter forms part of the ECHOES Project, which has received funding from the European Union's Horizon 2020 research and innovation program under grant agreement No. 770248.

L. Pozzi (✉)
Faculty of History, University of Warsaw, Warsaw, Poland
e-mail: l.pozzi2@uw.edu.pl

S. Lewis et al. (eds.), *Regions of Memory*, Palgrave Macmillan Memory Studies, https://doi.org/10.1007/978-3-030-93705-8_6

Public history is a lively field in China, where new projects are regularly opened (Li 2019). The relationship between history, memory, nation, and identity is particularly evident in state-financed projects, such as the multiplicity of oral history projects aimed at collecting the memories of the survivors of the War of Resistance against Japan (Li 2019). Schools, museums, documentaries, memorials, and heritage have become agents of memory that employ traumatic memories of colonialism and war to shape the identity of millions of Chinese people, intensifying at the same time nationalistic feelings among the population (Wang 2008). Since the 1990s, the authorities have been using the discourse of "national humiliation" (*guochi* 国耻) not only to legitimize their rule within the borders of the PRC, but also to shape Chinese foreign relations with Japan and the United States (Wang 2008, 800–2). The traumatic memories of humiliation, however, are of no use to support China's project of economic expansion in the Asia region. Therefore, to back their vision of China's leading role in Asia, the Chinese Communist Party (CCP) has been resurrecting collective memories related to the glorious past of China as the major crossroad of the Silk Road and the Maritime Silk Road. In this sense, the One Belt, One Road is also an exercise in collective memory at the transnational level.

Preservation of cultural heritage in the cities involved in the One Belt, One Road is one of the soft power strategies employed by the PRC to connect people from different countries to China's past (Winter 2019). In this framework, China's appropriation of the history of the Maritime Silk Road serves to promote a Sinocentric vision of the pre-colonial regional networks that connected countries in the Indian Ocean and South China Sea (Kwa 2016, 16).

This chapter analyzes how three museums in ex-colonial cities in Asia attempt to revive memories of China's central role in the Maritime Silk Road by reframing its heritage. It claims that exhibitions created in the framework of the One Belt, One Road aim not only at promoting China's political agenda, but also at overriding the history of European colonialism with new memories of China's glorious past. How is the expansion of China, defined by many as a neo-colonial power, changing the Asian region's memory of its colonial past? Is the PRC able to shape public history in and outside China's borders? Is the PRC able to propose an alternative to European colonialism by re-emerging pre-colonial memories of peaceful commercial exchanges in the Asia region?

Answers to these questions are gleaned through an analysis of the exhibitions of three museums: the Shanghai History Museum/Shanghai Revolution Museum, the Hong Kong Museum of History, and the Galle

National Museum in Sri Lanka. Based in municipalities that occupy different positions in the hierarchy of the One Belt, One Road initiative, each of these museums also reframes the heritage of the Maritime Silk Road in differentiated ways, providing us with relevant information on the PRC's political agenda and its gradations of influence over the Chinese-led construction of an Asian region of memory.

The Silk Road, the Maritime Silk Road, and the One Belt, One Road Project

In September 2013, Chinese President Xi Jinping launched a proposal for the construction of a new Silk Road by creating transportation corridors between China and Europe through Central Asia. One month later, during an official visit to Indonesia, Xi Jinping also launched the "21st Century Maritime Silk Road", a project aimed at opening new maritime economic routes linking deep water ports across the South China Sea, the Indian Ocean, and beyond. These two programs are collectively called the One Belt, One Road initiative, a project designed to promote China's political and economic position in Asia and in the world (Winter 2019, 9–18).

In the years following the announcement of this initiative, scholars have accused China of appropriating and manipulating the ideas of the Silk Road and the Maritime Silk Road for political purposes, using the past to serve the present-day foreign policy agenda (Kwa 2016, 1; Chan 2018, 160–61). One of the main criticisms raised against the PRC's heritage policies is regarding the historicity of the Maritime Silk Road itself and of China's supremacy in it. The idea of the Silk Road is a relatively recent historical construct coined by geographer Ferdinand von Richthofen (1833–1905) to describe the commercial routes that connected China to Europe as early as during the Han Dynasty (207 BCE–220 CE) (Waugh 2007). This concept was later popularized by Swedish geographer Sven Hedin (1865–1952), whose book entitled *The Silk Road* was published in 1938. More recently, scholars described the Silk Road not as a path, but as an ever-changing commercial, economic and cultural network that connected the Eurasian continent (Chan 2018, 161).

The concept of the Maritime Silk Road is even newer. Scholars began to employ it at the beginning of the twentieth century to describe cultural and commercial exchanges between ports in the Indian Ocean and the South China Sea (Chan 2018, 161). The history of the origins, development and consequences of these maritime networks has been at the center of historical debates from the beginning of the twentieth century, and historians are still studying the role of different civilizations in maintaining ties between ports in the region.[1] Mostly, the concept of the Maritime Silk Road is understood as the connections and interactions between three different trading worlds—the Arabian Sea, the Bay of Bengal, and the South China Sea—through the centuries (Kwa 2016, 19). Historians have studied these networks from different perspectives. For instance, in his analysis of the "Silk Road of the Seas", Lincoln Paine focuses on Muslim mariners' routes in the Indian Ocean (Paine 2013, 262–90), while Lo Jung-pang privileges the description of the maritime expansion of China (Lo 2012). Neither Lo—one of the major experts on Chinese maritime civilization—nor the historian of Chinese-Islamic relations Hyunhee Park (2012) defines the maritime networks that connected China to north-east Africa as a "Maritime Silk Road", demonstrating that this term is not indispensable in the analysis of the trading networks of the pre-modern Indian Ocean.

While historians tend to emphasize the transnational and multicultural nature of the maritime ties defined as the Maritime Silk Road, the cultural projects financed by the 21st Century Maritime Silk Road promote a Sinocentric vision of this historical regional network (Chan 2018, 160). According to scholars, President Xi Jinping has not only appropriated the idea of an ancient Maritime Silk Road as a metaphor of China's recent rise in the Asian region, but by promoting the centrality of China in this ancient network, he is also trying to "deepen the social memory of a regional integration in Southeast Asia and South Asia with China at his centre" (Kwa 2016, 17–18).

Heritage plays an essential role in the Chinese authorities' reconstruction of the regional memory of a Sinocentric Maritime Silk Road. One of the official goals of the One Belt, One Road initiative is to improve "people-to-people bonds", an aim that China is attempting to fulfil by using heritage domestically and internationally to reconstruct the history of the Maritime Silk Road (Winter 2016, 8). There are different examples of how the Chinese government supports the reframing of heritage to promote the One Belt, One Road initiative. In Guangzhou, local historians and policy makers worked together to prove that the city was one of the earlier ports from which Chinese mariners started their exploration of the South China Sea (Chan 2018, 165). Maritime archaeology also became a strategic tool to prove the far-reaching power of China overseas in the past. Chinese officials are financing underwater research and excavations to discover ancient Chinese vessels and ceramics in domestic as well as in

[1] For a concise overview of the historiography on the Maritime Silk Road see Kwa (2016).

international waters, for example off the coast of Sri Lanka (Adams 2013, Winter 2019, 119–25). The discoveries made during archaeological excavations are then handled by museums that contextualize vessels and potsherds to confirm the links connecting China to the rest of Asia with the aim of deepening the social memory of the Maritime Silk Road. Museums, however, adjust their narratives to the position their city occupies in the political and economic hierarchy of the One Belt, One Road initiative to promote slightly different views of China's role in the Asia region.

Creating a Past for Shanghai Before the Opium War: The Shanghai History Museum

The permanent exhibition of the Shanghai History Museum/Shanghai Revolution Museum (*Shanghai shi lishibowuguan, Shanghai geming lishi bowuguan* 上海市历史博物馆|上海革命历史博物; hereafter, the Shanghai History Museum) provides a thought-provoking example of how the PRC's institutions reframe heritage to fit the Maritime Silk Road narrative, offering visitors a new interpretation of the historical role of their city in national and international affairs.

After becoming an open port in the mid-nineteenth century, Shanghai started its ascent into what Jeffrey Wasserstrom calls a "global city", a neuralgic center for economy and culture at a global level (Wasserstrom 2009). Western historiography tends to overlook the history of pre-colonial Shanghai. For instance, in *A Short History of Shanghai* (first published 1927), author F.L Hawks Pott declares that while Shanghai existed before it became an open port, compared to many other places in China the city was "insignificant from [a] historical perspective" (Hawks Pott 2009, 2). This opinion is much contested in Chinese historiography, which sets the origin of the city during the Yuan Dynasty (1279–1368), more precisely in 1291 (Wasserstrom 2009, 2).

Despite work by local historians on the imperial history of the city, Shanghai remains mostly known for its economic and commercial growth during the colonial era. Between 1843 and 1949, a period referred to in China as "the century of humiliation", the city was governed by a system of concessions under the authority of several foreign powers, which also controlled other areas of the country. While Chinese historians strongly criticize foreign imperialism and highlight the abuses and injustices promoted by the colonial system in China, the idea that the prosperity of the city was due to its openness to modern technology and finance imported by foreigners proved long-lasting.

Public institutions in China advanced, and they are still advancing, the belief that while European colonialism was harmful to the Chinese nation, it nevertheless promoted the development of modernity in cities like Shanghai (Ifversen and Pozzi 2020, 157). For instance, exhibitions about the history of the city opened in the 1990s and the 2000s neglected the pre-colonial history of the city to focus instead on the period between 1843 and 1949 (Duan 2009, 32–41; Niu 2002; Pozzi 2021). These two exhibitions acknowledged the aggressive nature of foreign colonialism in China, nevertheless, they also glorified this epoch as a golden age for the economic and cultural development of the city, almost promoting a sense of nostalgia for the old Shanghai. Curators mention the lack of space in the building which hosted these exhibitions to explain the neglect for the pre-modern history of Shanghai (Duan 2009, 35), but the choice to promote the commercial growth of the city in the modern era also followed the contemporary political agenda to attract foreign investment and open the country to the world after years of political closure (Denton 2014, 88–94).

The current permanent exhibition of the Shanghai History Museum, which opened in spring 2018, overturns this tendency. Instead of focusing only on the colonial years, this exhibition analyzes the development of the city from its prehistoric origins to the present day, also prominently featuring its imperial history. Divided in two main galleries, the "Ancient Shanghai" gallery and the "Modern Shanghai" gallery, this exhibition employs the paradigm of the Maritime Silk Road to reframe the contemporary economic prosperity of the city as a prolongation of the commercial activities originated in the Tang Dynasty (618–907).

The Ancient Shanghai gallery aims at showing that Shanghai, at the time called Qinglong (青龙), was thriving before the arrival of western colonizers and that the prosperity of the metropolis in the twentieth century finds its origins in the vitality of the pre-colonial town. The captions and items in the exhibition show that the town of Qinglong was a one of the most prosperous ports of the Maritime Silk Road already during the Tang dynasty and that its economy continued to develop during the Song Dynasty (960–1279). Because of its financial success and strategic position, Shanghai was upgraded to the level of county during the Yuan dynasty (1271–1368). According to the bilingual captions (Chinese and English) that guide visitors through the exhibition, the "considerable amount of porcelain ware excavated in town indicates that it [Qinglong/Shanghai] was once an important port of transshipment along the Maritime Silk Road during the Tang and Song Dynasty." Porcelain wares excavated in the proximity of contemporary Shanghai are employed as evidence of the existence of the Maritime Silk Road and of the key-role that the city played in it.

Besides the archaeological remains, the exhibition employs innovative methods to attract the attention of visitors. A two-meter large screen placed in the middle of the main exhibition route plays a video entitled "The Maritime Silk Road", which restates the information provided in the exhibition captions in a trendier format (Fig. 1). The four-minute video, whose colorful images attracts the attention of visitors more than the fragments of porcelain, mixes animated recreations of the city's past with digital maps of ancient maritime routes. The video restates that "The port of Qinglong, the first in the region of Shanghai, was positioned on the estuary of the Wusong river. Since it was positioned on the river and on the sea, Qinglong became part of the Maritime Silk Road during the Tang and Song Dynasties." It also shows that during the Southern Song (1127–1279), the imperial court established in Qinglong an office for the

Fig. 1 Visitors of the Shanghai History Museum watch the video "The Maritime Silk Road" in November 2018. Picture by the author

supervision of shipment of goods.[2] Soon Shanghai became the seventh biggest port of the Maritime Silk Road. The video then continues describing the geographical and administrative changes that influenced the history of the city during the Yuan, Ming (1368–1644) and Qing (1644–1912) dynasties. Speaking about the colonial era, the video states that "after becoming an Open Port in 1843, the port of Shanghai and the city declared that Shanghai was going to become the world-famous hub port of the East". The short film concludes with images of the contemporary port claiming that nowadays Shanghai still supports "the spirit of the Silk Road" (*sichou zhi lu de jingshen* 丝绸之路的精神).

The characterization of Qinglong/Shanghai as one of the main ports of the Maritime Silk Road fulfils several aims. First, it counters the popular vision that, before becoming an Open Port, the city was irrelevant from historical perspective. Secondly, it shows that Shanghai was a "global city" maintaining international ties via the seaways before the arrival of the European colonial powers. Finally, it portrays the city as constantly evolving and growing in the framework of the Maritime Silk Road from the ancient past until its glorious present.

These claims are controversial. While China's commercial ties with other kingdoms in East Asia, Southeast Asia and South Asia are well-documented since antiquity (Lo 2012), to describe them in the framework of a well-defined and ever-expanding Maritime Silk Road comes close to historical revisionism. The fact that Shanghai is not even on the official list of the major Chinese cities applying for UNESCO protected status as part of the Maritime Silk Road (such as Nanjing, Penglai, Yangzhou, Ningbo, Fuzhou, Quanzhou, Zhangzhou, Guangzhou, and Beihai) supports the argument that the claims of the exhibition are dictated by current political contingencies (Wang and Cang 2015).[3] Furthermore, by linking the present commercial and financial success of the port of Shanghai to the older Maritime Silk Road, the exhibition trivializes the development of the city-port in the colonial era. The expansion of navigation in China, the ensuing development of the city's port, and

[2] The Song lost control of northern China to the Jurchen Jin Dynasty in1227. The Song withdrew to the south of the country.

[3] UNESCO supports the "Silk Roads Project", an online platform which "revives and extends these historic networks in digital space." China is one of the main developers of this initiative: https://en.unesco.org/silkroad/unesco-silk-road-online-platform (last accessed April 20, 2019). Since 2015, eight cities in China are trying to achieve UNESCO World Heritage Status as cities of the ancient Maritime Silk Road, see Wang and Cang (2015).

the construction of arsenals in the late nineteenth and early twentieth centuries are well studied, but they are hardly represented in the museum.

To conclude, the often-overviewed pre-colonial past of the city re-emerges in the permanent exhibition of the Shanghai History Museum, reframing it in the context of the 21st Century Maritime Silk Road project. They draw connections between the pre-modern and early modern Chinese maritime expansion and the contemporary role of Shanghai as one of Asia's most successful ports, effectively downplaying Shanghai's role as a global port in the colonial era.

The Maritime Silk Road and the Silk Road in the Hong Kong's Museums

The PRC's institutions are not the only ones employing the historiographical concepts of the Silk Road and Maritime Silk Road to reframe history and memory in the Asian region. On the contrary, in the attempt to shape visitors' understanding of China's international relations and to override the memory of the colonial past in the region, references to the Silk Road and to the Maritime Silk Road are also becoming more common in museums in the Hong Kong SAR.

Despite the "One Country, Two Systems" policy, political and economic interests between the PRC and its SAR are increasingly constraining curators' freedom to represent Hong Kong's past. In the early 2000s, the Home Affairs Bureau that supervises museums announced that Hong Kong "should recognize and return to its ancestors", while in 2009, the former Secretary of the Home Affairs Bureau and later Vice Chairman of the Silk Road Cities Alliance Zeng Decheng announced that Hong Kong's cultural policies should adhere to the functional role that "the Hong Kong people inherit Chinese culture" (Law 2013, 537).

The case of the Hong Kong Museum of History is exemplary of China's attempt to weaken the cultural and economic ties that connect the city to the United Kingdom, its former colonizer, and create a sense of belonging to the Chinese nation instead. The Hong Kong Museum of History opened in 1998, just one year after the hand-over of Hong Kong to the PRC. Besides hosting temporary exhibitions, in 2000 it also inaugurated its permanent exhibition, "The Hong Kong Story", which follows the history of the city from pre-history until the hand-over to the PRC in 1997.

According to the management, the museum's mission is "preserving and carrying forward the historical and cultural heritage of Hong Kong by collecting, restoring, sorting, and exhibiting archaeological [items], historical [events], folk customs, [and] natural and historical antiques in Hong Kong and surrounding areas" (Law 2013, 537). Nevertheless, the museum has received criticism from local and international scholars for its attempt to align the history of the city to those of the PRC. According to critics, the Folk Culture in Hong Kong gallery (Gallery 4) exhibits only Chinese traditional customs, while the western culture that has existed and exerted a profound influence on Hong Kong for more than 150 years is almost totally absent. Furthermore, Hong Kong's history as a British colony is reduced to a few small and fragmented segments (Law 2013, 539). Chen Yun, a highly influential cultural critic in Hong Kong, criticizes the exhibition as a "patriotic version" of the Hong Kong story designed simply to flatter the CCP regime (Law 2013, 538). Such criticisms, cast mostly by Hong Kong scholars, reveal the tension created by the expanding influence of the PRC over the city's cultural institutions and a fear that the Chinese authorities might try to elide the specificity of Hong Kong identity, which is still very much embedded in the colonial ties of the ex-colony with the United Kingdom.

The attempt to frame the history of the city in the Maritime Silk Road is another attempt to substitute Hong Kong's colonial legacy with more ancient ties to Chinese imperial history. The insertion of the Silk Road and the Maritime Silk Road narrative in the permanent exhibition of the Hong Kong Museum of History is much less visible in comparison to the case of the Shanghai History Museum. References to the Maritime Silk Road mostly appear in the explanatory captions of some objects discovered by archaeologist in the area of contemporary Hong Kong. In a similar fashion to the Shanghai History Museum, the permanent exhibition in Hong Kong claims that potsherds found in Penny's Bay prove that during the Ming Dynasty, "Hong Kong was a stop along the Maritime Silk Road to Guangzhou."

Certainly, the Maritime Silk Road paradigm is not as pervasive as in the Shanghai History Museum. The main reason for the difference approach of the two museums is that while the exhibition in Shanghai is brand new and therefore could fully develop a narrative fitting the government's agenda, "The Hong Kong Story" has not changed since its opening in 2000. Mentions of the Maritime Silk Road were added during the renovations of the permanent exhibitions in 2016, but no structural changes to the main narrative were applied.

Even if references to the Maritime Silk Road are minimal in the permanent exhibition, the institution has organized several events celebrating the Silk Road. For instance, between October and December 2016, the

Hong Kong History Museum organized a temporary exhibition entitled "Across the Oceans: The Local Connections and Global Dimensions of China's Maritime Silk Road", while between November 2017 and March 2018 it organized "Miles upon Miles: World Heritage along the Silk Road."

These exhibitions support the PRC's project to culturally integrate Hong Kong into the Mainland. The items displayed in "Across the Oceans: The Local Connections and Global Dimensions of China's Maritime Silk Road" came from the eight major cities on the Maritime Silk Road, and while Hong Kong is not on the list, the exhibition also included some objects found locally. According to the curators, "by using the cultural relics from eight Maritime Silk Road cities, as well as artefacts from Hong Kong, the exhibition elaborates upon the roles and functions each had as they grew and prospered on the Maritime Silk Road". Furthermore, artefacts from Hong Kong were added to the exhibition to create a sense of belonging and "to introduce the role played by the city in the history of the Maritime Silk Road." Like in the case of Shanghai, while Hong Kong is not on the list of the proposal for granting the Maritime Silk Road UNESCO World Heritage status, in the exhibition the metropolis is defined as a "Stop-over of the Maritime Silk Road."[4]

The temporary exhibitions organized by the Hong Kong Museum of History are just two of several events dedicated to the subject of the Silk Road organized in Hong Kong; many others have been organized by other institutions, such as the "Landscape Map of the Silk Road" (7 December 2018–20 February 2019) (Fig. 2). This exhibition, which opened in 2018 in the Museum of Science and Technology just in front of the Hong Kong Museum of History, showcased a 30-meter long and 0.6 meter wide Ming Dynasty silk handscroll representing a section of the Silk Road. The map presents the 211 cities between Jiayu Pass and Tianfanguo (nowadays Mecca), and it is a precious source to scholars interested in the history of mapping technology in pre-modern China (Lin 2018, 28–29).

From the artistic, scientific and historical perspectives, the value of this Ming Dynasty's handscroll is indisputable. As pointed out by Hyundee Park, maps can be used as texts which contain precious information about the geographical and cultural knowledge of a society in a given time (Park 2012, 14). While the exhibition "Landscape Map of the Silk Road" provided visitors with information on the development of mapping techniques in Chinese history and explained the unique qualities of this specific

[4] Hong Kong Museum of History, Past exhibition "Across the Oceans: the local Connections and Global Dimensions of China's Maritime Silk Road" https://hk.history.museum/en_US/web/mh/exhibition/2016_past_03.html (last Accessed April 15, 2019).

Fig. 2 Banners in front of the Hong Kong Museum of Science and Technology to advertise the exhibition "Landscape Map of the Silk Road" in November 2018. Picture by the author

map, the acquisition of this artwork, its re-naming, and its presentation in the 2018 exhibitions are a clear example of the reframing of historical heritage to fulfil political aims.

Firstly, the acquisition of this map by the newly established and not yet open Hong Kong Palace Museum relates to the current nationalistic political climate supported by the CCP. This scroll was kept in Yurikan Museum in Kyoto until 2002, when two collectors from Beijing bought it and brought it back to China. When the owners decided to sell the map, Dr. Shan Jixiang—the director of the newly established Hong Kong Palace Museum—decided to collect funds to acquire this piece for the museum. The fundraising proved unnecessary, as businessman Mr. Hui Wing-mau—a vocal supporter of the cultural reunification of Hong Kong with

China—acquired the map and donated it to the museum.[5] The choice of the acquisition of this specific map over other items is motivated by the strategic role of the Silk Road in contemporary Chinese politics. Furthermore, The Hong Kong Palace Museum, whose construction started in 2019, is in itself a political project: most of the items of the new museums will be borrowed from the original Palace Museum in Beijing, with the intent of creating stronger cultural links between the Chinese capital and the harbor of Hong Kong. Secondly, to make the association between the map and the ancient Silk Road clearer to the public, the name of the scroll, which according to historical sources was originally known as "Illustrations of the lands and people of the Western Regions" (Lin 2018, 28–29), was changed into "Landscape Map of the Silk Road." Lastly, the exhibition "Landscape Map of the Silk Road" while presenting the map as a precious source to understand Chinese cartography techniques, also states in its introduction that its presence in Hong Kong is a way to connect the city to China.

These examples show that museums in Hong Kong are systematically employing the heritage of the Maritime Silk Road and Silk Road to link the history of the city to the mainland. Their attempt to insert the history of the city into the Maritime Silk Road is more timid in comparison to the Shanghai History Museum, but the overwhelming number of exhibitions about China's domination over Asian aim at educating local visitors to the history of their "motherland", relegating the history of colonial Hong Kong to a secondary rank. Instead of criticizing the colonial experience of the city, cultural authorities have decided to downplay it by substituting it with a celebratory portrayal of the glorious expansion of the Chinese empire.

Chinese Civilization in the History of Southern Asia: The Galle National Museum

China's attempts to change the historical memory of the Asia region transgress the borders of the PRC and Hong Kong, spreading towards Southeast Asia and South Asia. Sri Lanka is an interesting example. Since the end of the civil conflict between the Government of Sri Lanka and the Liberation Tigers of Tamil Eelam in 2009, Sri Lanka has sought to attract foreign investors to rebuild the economy of the island, and while the

[5] Mr. Hui Wing-mau is the founder of the New Home Association (2015), which among other activities supports the exchange program "We are Family", which finances Hong Kong youth to visit China to explore its culture and history.

United States remained one of the main creditors, China has also started to offer financial assistance and loans to the Sri Lankan government (Mendis 2012). In May 2013, President Xi Jinping and President Mahinda Rajapaksa jointly announced the establishment of China-Sri Lanka strategic partnership, thereby including Sri Lanka in the 21st Century Maritime Silk Road initiative.[6] The PRC's interest in Sri Lanka is motivated by geopolitical and geo-economic reasons, as the island is a focal transit port within the international East-West shipping route in the Indian Ocean, and as such is a key-partner for China's geo-economic expansion (Mendis 2012). While the PRC's construction of a new port in Colombo and attracted the attention of scholars and politicians, China's investment in archaeological excavations in the island and the consequent construction of galleries in museums have been overlooked.

For the last two thousand years, Sri Lanka was a place where East-West shipping interacted and as such the whole coast of the island is of high interest for underwater archaeologists, who in the 1960s started researching shipwrecks in the area. One of the most successful projects was the excavation of the seventeenth-century wreck of the Dutch East India Company merchantman *Avondster*, initiated in 2001 with funding from the Netherlands Cultural Funds (Devendra and Muthucumarana 2013, 50–54). Ex-colonial rulers, mostly the Netherlands and Britain, have been the main financial supporters of these archaeological excavations, but in the early 2010s the PRC offered to help Sri Lanka to find Maritime Silk Road wrecks (Adams 2013). More recently, in October 2018, the Sri Lanka Central Cultural Fund signed a memorandum of understanding with the Shanghai Museum on a five-year archaeological cooperation program to excavate Northern Song Dynasty (960–1127) porcelain in the port of Jaffna. The archaeological team of the Shanghai Museum also hope to find more evidence of Chinese presence on the island in the fifteenth century (Xinhua Agency 2018). Some of the items retrieved during these underwater excavation missions—mostly Chinese porcelains and coins—entered the collections of local museums.

Differently than the Hong Kong case, the PRC has no direct legal rule over Sri Lanka. Still, the country is trying to support China's economic agenda, reframing archaeological findings as proofs of the long-lasting

[6] Ministry of Foreign Affairs of the People's Republic of China, Department of Asian Affairs, Sri Lanka (2013) https://www.fmprc.gov.cn/mfa_eng/wjb_663304/zzjg_663340/yzs_663350/gjlb_663354/2782_663558/. Last accessed April 18, 2019.

links between the two countries. One of the best examples of this tendency is the "Sri Lanka-China Historical Relationship" gallery opened in the Galle National Museum on September 10, 2013 to coincide with the visit to Sri Lanka of Liu Yunshan 刘云山, a member of the Standing Committee of the Political Bureau of the CCP's Central Committee.[7]

Opened in 1986, the permanent exhibition of the Galle National Museum includes pieces from the colonial history of the city, such as dresses and furniture of the Dutch and British colonists, and a collection of local masks and paraphernalia employed in religious rituals on the island. The "Sri Lanka-China Historical Relationship" gallery occupies a room at the very end of the building. Its ceiling is decorated with Chinese-inspired red and gold decorative motifs, and banners representing lions and dragons, traditional symbols of power in China, hang from the walls. The aim of the exhibition is to celebrate the 2000-year relationship between the two countries, which according to the introductory note located at the entrance of the gallery "existed mostly via the Sea Silk Road" and that had the distinctive feature of being always "peaceful and friendly."

The gallery displays coins and fragments of Chinese porcelain found in Sri Lanka as evidence of the commercial relations between the two countries since ancient times. These cultural relics, however, play a secondary role in the exhibition, whose main aim is the celebration of two Chinese historical figures who visited Sri Lanka in the pre-colonial era: Buddhist monk Fa Xian (法显; 337–422) and Admiral Zheng He (郑和; 1371–1433). Fa Xian—whose golden bust donated by Zhaoxianzhang Hanxuechen International Tour Management Association stands in the middle of the gallery—visited Sri Lanka on his return journey to China by sea after collecting Buddhist scriptures in India. His memoirs, which include references to religious rituals and the political situation in Sri Lanka in the fourth century, remain a precious historical source for researchers of pre-modern history.[8] The captions describing Fa Xian's travels underline the religious

[7] Ministry of Foreign Affairs of the People's Republic of China, Department of Asian Affairs, Sri Lanka (2013) https://www.fmprc.gov.cn/mfa_eng/wjb_663304/zzjg_663340/yzs_663350/gjlb_663354/2782_663558/ (last accessed April 18, 2019).

[8] See John Legge's translation of Fa Xian's *A Record of Buddhist Kingdoms: Being an Account of the Chinese Monk Fa-Hsien of his travels to India and Ceylon (399–404) in search of the Buddhist Books of Discipline* on Project Gutenberg EBook of Record of Buddhistic Kingdoms. http://www.gutenberg.org/files/2124/2124-h/2124-h.htm (last accessed April 20, 2019).

nature of the early contacts between Sri Lanka and China, claiming that the two countries were, and still are, Buddhist states. Given the centrality of Buddhism to Sinhala identity and culture, to present China as a custodian of Buddhist tradition is an extremely powerful statement aimed at drawing the two societies closer through religious feelings (Holt 2011, 334).

The gallery presents Fa Xian as a symbol of the religious connection between China and Sri Lanka, but the main hero of the exhibition is Admiral Zheng He. By order of Ming emperor Yongle 永乐 (1402–1427), this Muslim admiral led eight expeditions—seven official and one private—in the Indian Ocean between 1405 and 1433, going as far as the East Coast of Africa (Lo 2012, 333). Because of his role in Chinese maritime expansion, and the size of his fleet, Zheng He is celebrated as a hero in the PRC as well as in Singapore. In recent years, several museums and memorial halls dedicated to his endeavors have been established in China and abroad (Winter 2016, 10).

In the framework of the 21st Century Maritime Silk Road project, Zheng He is portrayed as an agent of peace and friendship between China and other nations. The reliability of this portrayal has been subject of scholars' scrutiny. During the Yuan Dynasty, commercial networks between China and the Islamic world through Southeast Asia and India flourished, but at the end of the Pax Mongolica, maritime trade in the Indian Ocean shrank (Park 2012, 167). The decrease of Chinese maritime trade after the fall of the Mongol rule was caused by the Ming court's ban on private sea travel (1371) and on travel relations with foreign countries (1381) (Park 2012, 167). While the Ming court attempted to circumscribe the expansion of private trade, it also sought to rebuild China's power over foreign countries through the re-establishment of a traditional tribute system. Zheng He's travels were aimed at rebuilding "China's commercial networks as an integral part of the state's tribute-base diplomatic order, as well as show off the power of their new empire" (Park 2012, 169). While Park claims that Zheng He's expeditions were peaceful and not comparable to European colonization project (Park 2012, 200), other scholars have shown that violence was often used during these diplomatic missions (Sen 2006). The case of Sri Lanka is one of the most revealing of the military nature of Zheng He's expeditions. During his second voyage in 1409, he erected a trilingual stele (in Chinese, Tamil, Persian) in Galle claiming Chinese suzerainty over the island (Lo 2012, 336). During his return to China after his third trip (1409–1411), Zheng He and his squadron were attacked by 50,000 Singhalese lead by Vira Vijaya Bahu VI, the king of Ceylon (reign 1397–1409). The Chinese marched inland and occupied the capital, and the king was taken as a prisoner to China, while a new more friendly Singhalese prince was chosen to rule the island (Lo 2012, 336–37).

Even though calling Zheng He's missions "peaceful" is anachronistic, the Sri Lanka-China Historical Relationship gallery celebrates the admiral's visits to the island as the peak of the two countries commercial relations. The plates' positions in the gallery claim that Zheng He "managed to maintain good rapport with the king and the administrators of the countries he visited" and mention the trilingual stone slab he installed in Galle during one of his trips as a sign of his benevolence towards the island.[9]

While captions describe Zheng He as a guardian of "peace" and "free trade" (another anachronistic concept) to prove China's interest in maintaining peaceful commercial ties in the country, the dioramas and models exhibited in the gallery endorse a different narrative: they compare Zheng He's missions to those of his contemporary European explorers, more precisely Christopher Columbus's (1451–1506) travel to the Americas. For instance, one of the display cases contains the replicas of Zheng He's ship and one of the "Ships of Christopher Columbus," accompanied by a detailed description of the measurements of the two. The larger size of Chinese vessels in the Ming Dynasty compared to European vessels is supported by historical sources (Lo 2012, 114; Park 2012, 333), but the aim of this display is not only to inform visitors of the historical details regarding Chinese navigation, but also to prove the superiority of Chinese technology and maritime skills over western equivalents. This display aims at undermining the relevance of western maritime expeditions to celebrate Zheng He's as a new Christopher Columbus, a comparison which has become common in scholarship despite the fact that Columbus navigated in unknown waters, while Zheng He followed routes already explored by many before him (Park 2012, 195–96). This narrative is strengthened by the content of the larger diorama in the gallery representing Zheng He and his collaborators landing on the coast of Sri Lanka: the men in the group, led by the admiral, wear clothes and hats of western features, and they are represented while pacing around their ship observing the foreign land (Fig. 3). The structure of this diorama has several elements in common with the celebratory images of Columbus arriving in Americas, such as the celebration of a more developed civilization coming to civilize the natives with their superior technical skills.

[9] The trilingual stele is now preserved in the Colombo National Museum, but copies are exhibited in the Galle National Museum and the National Maritime Museum in Galle.

Fig. 3 Diorama of the arrival of general Zheng He and his men in Sri Lanka in Galle National Museum in August 2018. Picture by the author

While the celebration of Chinese presence in Sri Lanka as peaceful and respectful of local traditions can be interpreted as an attempt to criticize the brutality of western colonialist powers, which exploited the natural resources of the island and tried to convert locals to Christianity, the celebration of Chinese maritime expansion and its comparison to European explorers hide a less peaceful subtext: the superiority of Chinese civilization in comparison with the local one. The not-so-subtle comparison of Zheng He to Columbus, whose endeavor is now considered as the beginning of centuries of suppression of civilizations living in the Americas and of western colonialism, contradicts the peaceful message promoted by the gallery. The celebration of Chinese civilization becomes a hymn to Chinese dominance not dissimilar to the claims of western superiority supported by colonial powers in Sri Lanka. Instead of celebrating the peaceful relations, the Sri Lanka-China Historical Relationship gallery not only promotes a revisionist history of the relationships between the two countries (defined as against historical evidence mirrors of the current nation-states), and instead of promoting equal foreign relations, it inadvertently presents China as a substitute to European colonial powers.

Conclusions

This chapter has analyzed how museums in ex-colonial cities in Asia attempt to construct memories of the Maritime Silk Road forging a new region of memory focused on China's peaceful rise as the main commercial player in the Indian Ocean and South China Sea. It showed how, in the context of the One Belt, One Road initiative, heritage of the maritime networks that connected China to the rest of Asia from the Tang dynasty onwards became a powerful tool to promote the PRC's political and economic ascent in the region.

The exhibitions taken into consideration share similar aims: to deepen the regional memory of China as the main political and economic power in the region and to challenge the historical and cultural ties that connect these cities with their European colonizers. In this sense, the One Belt, One Road project can be considered a transnational collective memory project as well as an economic one.

The narratives promoted in the three museums and their impacts, however, are adjusted to the history and legal framework of each city. The CCP can control the narrative in history museums in the PRC, and it is steadily extending its cultural and political influence on the Hong Kong's cultural scene. The Shanghai History Museum reframes the heritage of the Maritime Silk Road to demonstrate that Shanghai was a prosperous port before the Opium Wars, contradicting the popular view that the city was historically irrelevant before the arrival of the (western) colonizers. In the case of Hong Kong, the Maritime Silk Road is used to bind the city to the Mainland, foregrounding the cultural and historical legacies that connect Hong Kong to the PRC. By resurrecting the past glory of China, both museums downplay the impact of the colonial history of the city on its culture and economy.

The case of Sri Lanka is quite different, as the PRC's links to the country are mostly commercial. The museum in Galle demonstrates how the PRC employs heritage connected to its own history in foreign countries to promote its foreign policies and economic interests. Furthermore, it also shows that while museums in China do criticize the terrible consequences of western colonialism, this message gets lost in the PRC's self-promotion program. The Sri Lanka-China Historical Relationship gallery in the Galle National Museum does not criticize western colonialism of the island, instead it compares the deeds of Chinese explorers to those of contemporary European mariners, also implying that Chinese technology was superior. The exhibition revolves around Chinese travelers and Chinese culture, while the agency of local authorities and the accompanying violence is completely overlooked. If the reframing of the past is a reflection of present goals, the Sri Lanka-China Historical Relationship gallery betrays the neo-colonial ambitions of the PRC. Nevertheless, the PRC's political power in Sri Lanka is not comparable to its power to shape public history in Shanghai and Hong Kong. Despite its strong message, the Sri Lanka-China Historical Relationship gallery occupies only a small room of the whole museum, suggesting that while China uses its financial influence to intervene in Sri Lanka's cultural matters, there are limitations to how effectively it can promote its ideology on the island.

It is difficult, and perhaps too early, to assess the impact of these exhibitions on visitors in different cities and countries. The success of the PRC in changing the memory of the Asian region to a more Sinocentric vision of the past depends not only on the reframing of the heritage of the Maritime Silk Road, but also on its ability to create actual peaceful commercial and cultural connections among countries in the Asia region.

References

Adams, Jeff. 2013. The Role of Underwater Archaeology in Framing and Facilitating the Chinese National Strategic Agenda. In *Cultural Heritage Politics in China*, ed. Tami Blumenfield and Helaine Silverman, 261–82. New York: Springer.

Chan, Ying-kit. 2018. Cultural Heritage on China's 21st-Century Silk Road: The Case of Guangdong Province. *China Report* 54, no. 2: 159–74.

Denton, Kirk. 2014. *Exhibiting in the Past: Historical Memory and the Politics of Museums in Postsocialist China*. Honolulu: University of Hawai'i Press.

Devendra, Somasiri, and Rasika Muthucumarana. 2013. Maritime Archaeology and Sri Lanka: Globalization, Immigration, and Transformation in the Underwater Archaeological Record. *Historical Archaeology* 47, no. 1: 50–65.

Duan, Lian, ed. 2009. *Wangshi yu jiyi: Shanghai chu bowuguan, jinianguan koushu fangtang lü* [Former Events and Memory: A Record of the Interviews About Shanghai Museums and Memorials]. Shanghai: Shanghai cishu chubanshe.

Fa, Xian. 414. *A Record of Buddhist Kingdoms: Being an Account of the Chinese Monk Fa-Hsien of His Travels to India and Ceylon (399–404) in Search of the*

Buddhist Books of Discipline. Trans. James Legge. The Project Gutenberg EBook of Record of Buddhistic Kingdoms. http://www.gutenberg.org/files/2124/2124-h/2124-h.htm.

Hawks Pott, F.L. 2009. *A Short History of Shanghai: Being an Account of the Growth and Development of the International Settlement*. Beijing: China Intercontinental Press.

Holt, Clifford John. 2011. *The Sri Lanka Reader: History, Culture, Politics*. Durham and London: Duke University Press.

Ifversen, Jan, and Laura Pozzi. 2020. European Colonial Heritage in Shanghai: Conflicting Practices. *Heritage & Society* 13, no. 1–2: 143–63.

Kwa, Chong Guan. 2016. The Maritime Silk Road: History of an Idea. *Nalanda-Sriwijaya Centre, Working Paper* 23: 1–30.

Law, Kam-Yee. 2013. The Red Line Over European Colonialism: Comparison of the Macao Museum and Hong Kong Museum of History after Their Return to China. *International Journal of Heritage Studies* 20, no. 5: 534–55.

Li, Na. 2019. The Origin of Modern Public History in China. *History Workshop Journal* 88: 252–73.

Lin, Meicun. 2018. A National Treasure Coming Home: The Century-Long Homecoming Odyssey of the *Landscape Map of the Silk Road* in the Palace Museum Collection. In *Landscape of the Silk Road*, ed. the Hong Kong Museum of History, 28–37. Hong Kong: Hong Kong History Museum.

Lo, Jung-pang. 2012. *China as a Sea Power 1127–1368: A Preliminary Survey of the Maritime Expansion and Naval Exploits of the Chinese People During the Southern Song and Yuan Periods*, ed. Bruce A. Elleman. Hong Kong: Hong Kong University Press.

Mendis, Patrik. 2012. The Colombo Centric New Silk Road. *Economic and Political Weekly* 47, no. 9: 69–76.

Niu, Weiping, ed. 2002. *Bainian huiwei: Shanghai Lishi fazhan chenlieguan xunli* [The Pleasant Flavour of the Last One Hundred Years: A Tour of the "Exhibition of the Development of Shanghai Modern History"]. Shanghai: Shanghai yuandong chubanshe.

Paine, Lincoln. 2013. *The Sea and Civilization: A Maritime History of the World*. New York: Vintage Books.

Park, Hyunhee. 2012. *Mapping the Chinese and Islamic Worlds: Cross-Cultural Exchange in Pre-Modern Asia*. Cambridge and New York: Cambridge University Press.

Pozzi, Laura. 2021. Local Museum, National History: Curating Shanghai's History in the Context of a Changing China (1994–2018). *International Journal of Heritage Studies* 27, no. 4: 407–22.

Sen, Tansen. 2006. The Formation of Chinese Maritime Networks to Southern Asia 1200–1450. *Journal of the Economic and Social History of the Orient* 49, no. 4: 421–53.

Wang, Xin, and Wei Cang. 2015. Cities Unite to Seek Historic Status for Maritime Silk Road. *China Daily*, The State Council of the People's Republic of China. July 6, 2015. http://english.cndy.org/news/top_news/2015/07/06/content_281475141388714.htm.

Wang, Zheng. 2008. National Humiliation, History Education, and the Politics of Memory: Patriotic Education Campaign in China. *International Studies Quarterly* 52, no. 4: 783–806.

Wasserstrom, Jeffrey N. 2009. *Global Shanghai, 1850–2010: A History in Fragments*. New York: Routledge.

Waugh, Daniel. 2007. Richthofen's 'Silk Roads': Toward the Archaeology of a Concept. *The Silk Road* 5, no. 1: 1–10.

Winter, Tim. 2016. Heritage Diplomacy Along the One Belt, One Road. *The Newsletter* 74: 8–10.

———. 2019. *Geocultural Power: China's Quest to Revive the Silk Road for the Twenty-First Century*. Chicago: The University of Chicago Press.

Xinhua Agency. 2018. China, Sri Lanka Join hands in Maritime Silk Road Archaeological Research. *Global Times*, September 12. http://www.globaltimes.cn/content/1119309.shtml

PART III

Cultural Regions of Memory

Articulations of Memory: Mediation and the Making of Mnemo-Regions

Ann Rigney

Postnational Memory Studies

There is evidence of a growing desire among memory scholars to go beyond "methodological nationalism" (Amelina et al. 2012), reflected in the appearance of a flock of related terms: "cosmopolitan memory" (Levy and Sznaider 2002), "memory in the global age" (Assmann and Conrad 2010), "travelling memory" (Erll 2011), "transcultural memory" (Crownshaw 2014; Bond and Rapson 2014), "transnational memory" (De Cesari and Rigney 2014), and "memory in the globital age" (Reading 2016). These concepts overlap and sometimes compete, each facilitating a different type of analytic work. But they all share a common concern with thinking about new frameworks and new sites for studying the production of memory in an increasingly entangled world of globalized communication, mass migration, and planetary climate change. Together, they represent a concerted attempt to conceptualize memory in what can be called

A. Rigney (✉)
Utrecht University, Utrecht, The Netherlands
e-mail: a.rigney@uu.nl

S. Lewis et al. (eds.), *Regions of Memory*, Palgrave Macmillan Memory Studies, https://doi.org/10.1007/978-3-030-93705-8_7

postnational terms, with "postnational" being understood here by analogy with the term postcolonial, as referencing a world that recognizes the enduring legacy of nationalism, while also relativizing the current importance of the nation state in organizing social, cultural, and political life. The concept of region, central to the present volume, belongs in this postnational movement in memory studies.

By and large critiques of methodological nationalism have involved broadening the framework in which memory production is considered. In a milestone essay called "Travelling Memory," Astrid Erll argued that memory is "not bound" by any national container and, instead, is "continually moving across and beyond [such] territorial and social borders" (Erll 2011). Her use of the term "not bound" echoed Levy and Sznaider's article "Memory Unbound" (2002) and was subsequently echoed in the (sub-)titles of later essay collections, including: *Memory Unbound* (Bond et al. 2017), and *Anne Frank Unbound* (Kirshenblatt-Gimblett and Shandler 2012).

The concept of the unbound represents an important first step away from the nation state as a self-evident framework for memory and identity. In particular, it has helped yield a growing body of insights into how global entanglements are explored in the arts. However, as I shall argue here in line with De Cesari and Rigney (2014), the "unbounded" character of memory should not be overemphasized, lest a critique directed at relativizing the national as a framework for memory should end up creating the illusion of a brave new world with no borders at all; with no outside and no outsiders. The contemporary world is not a seamless space where everything flows smoothly across borders; nor is it a level playing field where all experiences have equal weight or voices equal power. Arjun Appadurai has referred in this regard to the "bumps and potholes" (2003) that interrupt the flow and uptake of information and affect across the world. With this in mind, the present article approaches memory production as a multi-scalar and multi-sited process operating across uneven topographies. Material and imagined borders between "insiders" and "outsiders" still exist even if they no longer coincide with the traditional borders of nation states. Indeed, the number of borders or interfaces has arguably grown rather than diminished as a result of our entanglements. If Europe, for example, has been about "integration" and the disappearance of border controls, it has also generated a proliferation of new borders operating at different scales and marking the outer reaches of what is considered "European" (see Rigney 2012b). In order to study this complexity, memory studies should not dismiss the national as outdated. Not only because nationalism is proving to be a hugely resilient force in society, but more importantly here, because contemporary memory studies may also learn from the history of nationalism as an earlier phase in the "upscaling" of society.

Critiques of the nation state as the "natural" framework for collective identity and memory actually predate the publications mentioned above. Arguably, reflexivity regarding the phenomenon of the nation state has been at the heart of memory studies since its emergence as an interdiscipline in the 1980s. Pierre Nora's influential *Lieux de mémoire* project (1984–1992) was in essence a nationalist project in that it buttresses a canon of French identity. But the very fact that it thematized the nation state as a framework for memory indicates that the latter had lost its self-evident primacy. Nora's work is often cited as a key point of reference in genealogies of memory studies. What has less often been noted is that it appeared around the same time as Hobsbawm and Ranger's *The Invention of Tradition* (1983), David Lowenthal's *The Past is a Foreign Country* (1985) and, last but by no means least, Benedict Anderson's *Imagined Communities* (1991 [1983]). What these works have in common is that they reflect on the nation as a "natural" mnemonic frame by exploring the history of its construction.

In his influential study of "imagined communities", Anderson invoked Ernest Renan's famous "Qu'est-ce qu'une nation?" ([1882] 1947–1961)) in order to argue that remembrance and forgetting are key aspects of nation-building. Nationalism, Anderson argued, was "a new way of linking fraternity, power and time meaningfully together" (1991, 36). It involved a voluntary identification—or what Edward Said around the same time was calling "affiliation" (Said 1983)—with people with whom one had no "natural" or traditional connections. This voluntary affiliation yielded a fictive kinship (what Anderson called "imaginings of fraternity"; Anderson 1991, 203) based on the sense of a shared history; this involved overlooking genealogical differences and forgetting divisive memories. Learning to "live with strangers", to echo Kwame Anthony Appiah's definition of cosmopolitanism (2006), was arguably already part of nation-building and its production of fraternal solidarity across greater distances.

In the case of nation-building, however, "linking fraternity, power, and time" did not entail a permanent openness to strangers. In contrast to cosmopolitanism, nationalism involved both mechanisms of bonding and, once the common history had been established, mechanisms of exclusion:

it depended on the idea that some strangers were brothers because they shared a common history and culture, while others were constitutionally outsiders. Nations thus came into being by generating a demarcation line between "us" and "other", between home and abroad, the domestic and the foreign. As Anderson noted, nations are by definition "limited" or, indeed, bounded:

> The nation is imagined as limited because even the largest of them, encompassing perhaps a billion living human beings, has finite, if elastic boundaries, beyond which lie other nations. No nation imagines itself coterminous with mankind. (Anderson 1991, 7)

In what follows, I propose to take a leaf out of the book of nationalism studies in recognizing people's capacity to voluntarily affiliate with strangers, but also the inevitable "boundaries" of all imagined communities; indeed of identity itself, defined as the subjective sense of belonging to "a bounded, distinctive, solidary group" (Brubaker and Cooper 2000, 34). Although the boundaries of imagined fraternities may be elastically extended, they never become coterminous with mankind, not necessarily because of deliberate mechanisms of exclusion (as in nationalism), but because of the limits of communication and of our ability to imagine the lives of others.

Having arrived at this point in my argument, the concept of region presents a welcome alternative to the nation as a conceptual tool for thinking about the role of borders in memory production. Like a nation, but without its exclusionary logic, a region designates a structure *in between* the global and the local. It is bounded, by definition, but its borders do not coincide with the nation and, crucially, they are also subject to being redefined. In the terms of the Merriam-Webster Dictionary, a region is "an indefinite area of the world or universe"; in the terms of the Oxford dictionary, a region is "an area, especially part of a country or the world having definable characteristics but not always fixed boundaries".

In what follows, I shall adapt this notion of region to study the circulation but also the boundaries of memory in a postnational world. More specifically, I explore the role of the media in defining but also shifting those boundaries by creating the imaginative conditions for new affiliations with strangers and hence virtual mnemo-regions that transcend traditional solidarities.

Mediation and Cultural Borders

A key assumption in cultural memory studies is that mediated acts of remembrance are performative in the strong sense of speech act theory (Austin 1962): they not only reflect the world but also effect change in it, by giving meaning and affective force to the memory of events. In this sense, mediated acts of remembrance are agents of "worlding" (Anderson and Harrison 2010).[1] The very fact of sharing narratives through communicative acts specifically helps to shape social relations by providing common points of reference and a common language. These points bear emphasizing since the argument that follows rests on the assumption that media do not just reflect social relations but, in certain circumstances, also have the power to change them by changing lines of communication and patterns of affiliation.

In highlighting the role of media in producing collective identities, cultural memory studies shares a common focus with nationalism studies. The invention of nations in the nineteenth century (Leerssen 2018) offers a prime example of the performative power of cultural practices, including mnemonic practices, in forging an imagined fraternity between increasingly nationalized subjects and in promoting the desire for congruence between cultural identity, state formation, and national territory. Anderson (1991) notably attributed the making of nations to the role of modern media in creating imagined communities at a scale transcending that of local, face-to-face contacts. With a predigital, indeed pre-audiovisual view of the world, Anderson restricted himself to print media (newspapers and the novel), focusing on how they operated within specifically national frameworks especially in the period before World War One. People now live in more complex media ecologies, but the basic point is still relevant: namely, that the media, as carriers of memory, have the power to create new alliances and new imagined communities. In today's digital world, this power clearly operates at scales both below and beyond the national frame.

Contemporary media do not stop at a country's border no more than the weather does. Nor did they do so, *pace* Anderson, in the age of print. Like the flight paths that are displayed in in-flight magazines, communication networks traverse and transcend national territories and produce

[1] Anderson and Harrison (2010, 8): "the term 'world' does not refer to an extant thing but rather the context or background against which particular things show up and take on significance: a mobile but more or less stable ensemble of practices, involvements, relations, capacities, tendencies and affordances".

distinct constellations. The circulation of information brings a world of connections into being and, through the exposure to common narratives, imagined communities shaped along other than national lines and not necessarily tied to a particular territory. Communication networks transcend physical space and create new, de-territorialized relations of distance and proximity (Anderson and Harrison's concept of "non-representational" geography is useful here; 2010). In keeping with the key concept of the present collection, we can speak here of "mnemo-regions" in the sense of virtual, de-territorialized mnemonic formations that are coterminous neither with national frameworks nor with the planet as a whole. As the idea of "region" indicates, these formations are bounded, although the location of that boundary is not defined a priori but is continuously emerging as one of the outcomes of communication and connectivity itself.

Knowing more about communication networks is a precondition for understanding how memory is produced transnationally today. Who is talking to whom and which narratives reach which destinations? Addressing these questions entails studying the way narratives "travel" but also how their movement may be contained within echo chambers of the likeminded or blocked by cultural borders. Unlike the borders of nation states which are linked to clearly-defined jurisdictions, cultural borders are never clear cut, open or shut. Instead, as the anthropologist Claude Lévi-Strauss once argued, they take the form of a series of thresholds each marking a gradual decrease in mutual comprehension because of differences in language, values, and habitus (Lévi-Strauss 1974). This suggests that, in certain circumstances, mediation can help to shift cultural borders by *creating new grounds* for mutual comprehension between hitherto disconnected parties who then form, if only temporarily, a new public (Warner 2002).

The following analysis builds on this hypothesis by considering the various ways in which mediated memories create new interfaces between hitherto disconnected groups. This research question resonates with Astrid Erll's recent analysis of "relationality" (Erll 2017, 2018); specifically, of the various ways in which the creative arts reveal underlying entanglements between the memories of distinctive groups. My emphasis will be less on the uncovering of hitherto overlooked connections, however, and more on the forging of new interfaces between as yet unconnected groups through acts of mediation. A key concept in this analysis is that of articulation, building on De Cesari and Rigney (2014).

As Stuart Hall already pointed out (in Grossberg 1986), the word "articulation" means, on the one hand, "speaking clearly and coherently" and, on the other hand, "connecting" (as in joints and trucks). Its value as an analytic term is that can help us bring together within the same frame both the composition and the circulation of cultural artefacts. Applied to memory production, articulation entails mediated acts of remembrance that "perform" connections between people. When memory is given expression in a cultural carrier (novel, film, testimony, documentary...) that goes into circulation, new connections between individuals, between groups, and between "us" and "strangers" are made possible. Acts of articulation thus enable "unscripted linkages" (Rothberg 2014, 133), affiliations that are not based on tradition, but are newly constituted (as national identities once were). Alison Landsberg (2004) has influentially argued that experiential modes of narration in particular can produce "prosthetic memory" whereby people engage imaginatively and affectively with a history into which they have not been socialized, but that now becomes part of their memory and identity thanks to their exposure to the narrative. Landsberg is arguably over-optimistic regarding the empathy-inducing powers she ascribes to experiential narrative. But her basic point is nevertheless well taken; namely, that cultural carriers, fictional narratives in particular, can form a bridge between mnemonic communities; in our terms, articulate them.

Four Types of Articulation

What follows is an initial typology of the different ways in which historically distinct memory communities can be "articulated", that is, brought into conversation through mediated narrative practices. Four types of articulation, touching on both the composition of narratives and the manner of their circulation, will be briefly discussed. The examples offered to illustrate these types are wide-ranging, designed to show the complexity of the phenomenon but also to serve as a reminder that the work of articulation is not just a by-product of the postnational, but has long been a part of mediatized community-building. In the last section of this article, I will go on to consider the work of articulation within specifically digital contexts.

Integrative narration involves using a plot structure to reconcile two opposed groups as part of a single "birth of the nation" story with a happy end. This was the classic narrative model for nation-building and linked to the genre of the historical novel as this emerged at the beginning of the nineteenth century. It is exemplified in the novel *Ivanhoe* (1819), the international bestseller by Walter Scott about the struggle between the native Anglo-Saxon population and the Norman colonists in the century after the Norman Conquest of Britain (c.f. Rigney 2012a). Like Scott's other novels, it offered a way of imagining the birth of the nation as the overcoming of ethnic differences or, to use our terms here, the articulation of two groups. Renan's claim that all the nations of Europe originally originated in two "races" (Renan 1947–1961) was arguably inspired by this integrative narrative model whose emergence can be seen as a response to one of the major challenges of nation-building; namely, how to imagine a single nation on the basis of differences? More recent examples of this narrative model can still be identified (for example, in the American television series *North and South* (1985–1994), based on the novelistic trilogy by John Jakes (1982)). In the context of contemporary migration, other variations have come on the scene. These work less on the principle of two-becoming-one in a single plot than on showing the legitimate co-existence within the same narrative frame of different groups with their own histories. More could be said on the rich array of narrative models that are still being invented in response to changing conditions in the life world and changing ideas about citizenship. Suffice it here to mention, as one example among many possible ones, the comic novel the *Clash of Civilizations over an Elevator in Piazza Vittorio* (2008) by Italian-Algerian writer Amara Lakhous, which thematizes cultural and linguistic differences, and divergent life narratives among people in contemporary Rome. In this case, the convivial sharing of city space is the unifying principle rather than the unfolding of events over time, but here too, narration is being used in an integrative way.

Multidirectional narration involves bringing into conversation within the framework of the same cultural text historical experiences hitherto unconnected in public perception. The experiences are not so much connected at the level of the plot as through cross-referencing and analogies. The term "multidirectionality" is drawn here from Michael Rothberg's seminal work (2009) which shows how the memory of the Holocaust provided a model for articulating the memory of colonization in the 1950s, and vice versa, how colonial memory has helped in giving expression to the memory of the Holocaust. To take one of Rothberg's examples: Caryl Phillips' novel *The Nature of Blood* interweaves two story lines—one is about a Holocaust survivor in the twentieth century; the other about Othello in Renaissance Venice—into mutually illuminating juxtaposition or "multidirectional montage" (Erll 2017). Phillips jumps imaginatively through time and space and, in the process, invites his reader to make connections between two forms of racism without integrating the different subplots into a single story.

Hetero-remembrance involves narrating the historical experiences of an outgroup or the "lives of others". It is not a feature of narrative composition but of positionality. Most theorizing in memory studies has been focused on what might be called "auto-remembrance": the ways in which individuals and groups negotiate and give expression to their own memory, be this as a private person or as members of a community. As a result, the ubiquity of cross-cultural remembrance—what I call here "hetero-remembrance"—has been overlooked. More than has been recognized hitherto, historians, writers, directors, curators, and other memory workers express an interest in the experience of groups with whom they are not traditionally affiliated by language, history, or the sharing of a particular territory. The work of Lydia Koidula, penname for the German-educated Estonian writer Lydia Jannsen, offers a case in point. As Piret Peiker (2015) explains, Koidula's historical fiction *Juudit, or the Last Maroons of Jamaica* (1870) narrates in Estonian the story of runaway slaves fighting the British in the Caribbean and thus very far from the Baltic and the Russian Empire. Koidula's story was read in Estonia as being about "the life of others" to be sure, but also as a story that resonated (Armstrong and Crage 2006) with that subaltern nation's struggle for independence vis-à-vis both the German elite and the Russian empire. An experiential narrative about a group located elsewhere was thus made imaginatively available to a domestic audience for whom it also became a multidirectional projection screen for anti-colonialist concerns closer to home. A similar point might be made about the ways in which the French revolutionary leader Danton has been depicted outside of France from Georg Büchner's *Dantons Tod* (1835) to the movie *Danton* (1983), by Polish director Andrzej Wajda; or the Paris Commune, as depicted for example in Bertolt Brecht's *Tage der Commune* (1949), a German play about the Paris Commune which was recently reproduced again, as performance and as video installation, in New York within the context of the Occupy Wall Street movement.[2] Hetero-remembrance transcends—effectively suspends—linguistic, cultural, and mnemonic borders. In doing so, it makes

[2] *Days of the Commune* was directed as street theatre by Zoe Beloff, recordings of which were later made available on vimeo (www.vimeo.com/55209376); the project also became the subject of an installation at the Slought Foundation, Philadelphia in May 2011; see http://daysofthecommune.com/.

possible different types of interaction between self and other, and promotes some degree of identification with actors at a different location. In doing so, our examples suggest, it may be a way of holding up a critical mirror to the home country. In other instances, a selective interest in the "lives of others" may be a way of keeping them "in their place" within a traditional hierarchy (hetero-remembrance in the recent British/US mini-series *Chernobyl* (dir. J. Renck 2019), for example, shades over into an exoticization of the former USSR). Whatever the particular effect intended or achieved, hetero-remembrance suspends the boundaries between "home" and "abroad", and if only in a partial or temporary way, relocates the imagined boundaries between "us" and the "other".

Translation, as the name indicates, involves transposing existing narratives across linguistic borders and "translation zones" (Apter 2006). Translating novels and subtitling movies is the converse of hetero-remembrance in that it involves making accessible narratives that originated elsewhere. Translation from one language to another has largely been overlooked in studies of transnational memory work, though in many cases it is a necessary condition for the "travelling" of memory (Erll 2011). Non-fictional historical works are regularly translated, but it would seem that imaginative works circulate more easily and more frequently across linguistic borders. As I have argued elsewhere, the arts do not require a prior interest in a particular topic, but are instead capable of generating that interest through the way in which particular narratives are composed and offer aesthetic rewards (see Rigney 2009, 2021). Javier Cercas's novel on the Spanish civil war *Soldiers of Salamis* (2001), for example, has been translated to date into 15 languages, including Japanese. Orhan Pamuk's novel *Snow* (2002) about the unrest relating to the history of Kurds and Armenians in Eastern Anatolia has been translated into 19 languages, including Japanese and Chinese. A similar point can be made about the subtitling (and foreign distribution) of films dealing with historical events. Among many possible examples, mention can be made here of Andrzej Wajda's *Katyn* (2007), distributed with subtitles in 30 European countries; Paul Greengrass's *Bloody Sunday* (2002) distributed in 33 countries; and Emanuele Crialese's *Terraferma* (2011) distributed with subtitles in 28 countries.[3]

[3] Data collected in November 2019. Information on the distribution of European films has been found at: https://eacea.ec.europa.eu/creative-europe/funding/distribution-selective-scheme-support-for-transnational-distribution-european-films_en. A more detailed study

While the very existence of translations and subtitled versions indicates that "foreign" narratives are of some domestic interest, the variation in the number of translations shows that memories move across uneven ground. Translations are facilitated or impeded according to priorities in the target culture or, conversely, the extent of the efforts made in the source culture to promote particular works abroad. For it is clear that not all novels and films travel to the same extent: because of a lack of institutional support, a lack of aesthetic appeal, a lack of local resonance in the subject-matter, or a combination of all of these factors. To date, Dubravka Ugrešić's *The Museum of Unconditional Surrender* (1998), originally written in Serbo-Croatian, on the legacy of the former-Yugoslavia has only been translated 4 times, while Assia Djebar's novel on the Algerian civil war *Children of the New World* (1962), originally written in French, has only been translated into English. Finally, the film *Terraferma* about illegal immigrants trying to survive in Sicily has been widely distributed in Western Europe (as well as Brazil, Israel, Korea, and Taiwan), but it has not yet travelled widely in Eastern Europe, with the exception of the Czech Republic, Hungary, and Ukraine.

Mapping and explaining the unequal distribution of such movies and translations is a desideratum in memory studies.[4] In this regard, scholars would do well to take a leaf out of the book of transnational poetics, translation studies, and world literature studies where the question is asked how receiving cultures "negotiate, accept, indigenize, resist, and transform foreign influences" (Ramazani 2009, xiii).[5] That "foreign influences" may be appropriated for local purposes is borne out by the reception of Paul Greengrass's movie *Bloody Sunday* (2002), which was given a public screening in Bozen in South Tyrol in 2009 on the occasion of the bicentenary of the death of local resistance hero Andreas Hofer. The story of

would need to extend the scope of the analysis to include distribution patterns elsewhere in the world. Information on translations has been gained from UNESCO's Index Translationum (http://www.unesco.org/xtrans/) supplemented by online searches for particular writers.

[4] Happily, after this article was originally written, a project called *Translating Memories*, directed by Eneken Laanes at Tallinn University, was started thanks to funding from the European Research Council: www.translatingmemories.tlu.ee.

[5] The study of world literature and translation studies have focused largely on the tensions between global culture (largely English-language) and linguistic diversity, and on the desirability or not of maintaining "untranslatability" as a form of resistance to cultural homogenization; see for example, Damrosch (2003) and Apter (2006). Translation studies too has focused on the tension between domestication and "othering"; see for example Venuti (2013).

the Irish Bloody Sunday apparently resonated with the local narrative of struggle against foreign oppressors (also presumably with the South-Tyrolean version of "Bloody Sunday" in 1921 when Italian-led fascist gangs attacked locals).[6] In order to understand the ways stories originating in one culture are appropriated and interpreted in others, more attention should be paid in the future both to public practices of reception and to the (semi-)private practices of reception evidenced, for example, on online forums such as Goodreads. The idea that new stories need to resonate with existing ones in order to take root, as Armstrong and Crage have argued (2006), also implies the need for future research into the multifarious ways in which "foreign" stories map onto domestic ones.

But hopefully enough has already been shown here to support the point that narratives, both in their composition and their transposition to other places, have the potential to change traditional patterns of affiliation and to *create* new relations between hitherto unconnected mnemonic communities. The composition, circulation and local uptake of narratives create "flight paths" that transcend linguistic, cultural, and national borders and extend the boundaries of imagined communities into new mnemo-regions.

My examples so far have been from traditional broadcast media and have specifically referred to experiential narratives produced by creative writers and filmmakers. In my final section, I reflect on how these basic insights apply to today's digital "culture of connectivity" (Van Dijck 2012).

Wiki-memory

If there was a natural affinity between newspapers, novels, and nations, as Anderson argued, what sort of social formations are sustained by the new media ecologies? And do they too carry out acts of articulation? Not surprisingly, there is a fast-growing literature on the impact of digitalization on cultures of memory and the collaborative production of memory narratives, which it is beyond the scope of this article to discuss in detail. Suffice it here to flag two general issues. Firstly, as several important studies have shown, digitalization has generated multiple new possibilities for storing and sharing information independent of experts and state institutions (Garde-Hansen et al. 2009; Reading 2016), for building more

[6] On the Tyrolean Bloody Sunday, Steininger (2003) and Thaler and Mumelter (2011). I am grateful to Susanne Knittel for drawing my attention to this case; for details see Rigney (2016).

personalized archives (Van Dijck 2007), and for the online remediation of textual and audiovisual materials (Jenkins 2006), which makes them accessible to new generations and new publics irrespective of their location. Secondly, and more challenging, theorists of digitally-born communication have argued that the traditional narrative forms of memory-making exemplified in the previous section of this paper have become subject in the new media ecologies to a "connective turn" (Hoskins 2011). There has been a paradigm shift, Hoskins argues, whereby the recollection of the past as experiential story has become less important to collective identity than the sense of being connected or networked in the present; this connectivity is not dependent on empathic engagement with the lives of others, but rather on the algorithms that link the users of social media and hence the technologies that "engineer" their sociality (Van Dijck 2007, 169).

Since digital culture is evolving as we write and encompasses many different practices and technologies, it is impossible to generalize. For this reason, I have decided to focus specifically on Wikipedia, since this online encyclopedia, unlike other social media platforms like Facebook, is more obviously linked to collective memory and the shaping of imagined communities, and thus offers a good observation point for the ongoing transformation of memory culture. Wikipedia has been identified as a distinctive "technology of memory" (Plate and Smelik 2009), and also as a "global memory place" (Pentzold 2009), its scope and its multilingualism exemplifying postnational memory production. Its design means that it is also an important site of "convergence culture" (Jenkins 2006) where different cultural practices are brought together on one platform. Most importantly here, it is a generator *par excellence* of articulations which uses, but also goes beyond, the forms outlined above. As we will see, Wikipedia uses the possibilities offered by digital media to continue the work of articulation across a topography that, albeit free of national frames and unprecedently global in its scale, nevertheless remains uneven and differentiated.

In the course of just a decade, Wikipedia has become a key site of transnational memory production (Pentzold 2009; Pentzold et al. 2017; Ferron and Massa 2014; Rosenzweig 2006). There are some signs that activity has leveled off since around 2014 and that a certain editorial fatigue has set in, but it still remains a hugely important medium of memory with a unique set of affordances: an interactive media technology, a template for ordering information, and a system of governance that allows both for user generated content and regulation on the part of editors (Van Dijck 2012). The governance model is aimed at achieving accuracy, neutrality, global reach, and consensus-building in an international arena (accessing the "edit history" reveals the level and intensity of controversy on particular topics).[7]

Wikipedia's linguistic range is impressive, with no fewer than 301 languages represented as of July 2019. The English-language Wikipedia is by far the largest and seems to serve constituencies both inside and outside the English-speaking world, hence giving it both a global and a regional-specific dimension. A comparative study of the different articles on the Srebrenica massacre in Serbian, Croatian, Dutch, and English, led by Richard Rogers at the University of Amsterdam (2013), concluded that the same editors act differently on the English-language site (targeting an international public and subject to greater control) than on the site in their local language (where articles were more colored by interpretation). Since machine translation is only permitted in a limited way,[8] the appearance of new articles (be this in the form of a translation or a new text) can be taken as symptoms of interest in a certain topic and a desire to make the memory available to a new linguistic community. Since Wikipedia is a non-aesthetic medium without the attractions of experiential narrative, the uptake of these articles will depend either on the prior interest of reading subjects or on serendipity, that is, users happening upon an article by following other links, which the hyper-linked interface encourages them to do.

Despite its potential to accommodate all interests, attention in Wikipedia is not spread evenly over languages and topics. National minority languages like Breton and transnational languages like Tatar are "over"-represented through multiple articles, with Wikipedia apparently offering a new international platform for non-state languages and the cultural identities they sustain. In contrast, African languages and coverage of Africa-related topics are under-represented leaving that continent largely outside of the over-lapping mnemo-regions being produced by this internationally used technology.[9]

[7] Additional digital tools can give more information about patterns in editing and moderation: http://contropedia.net/.

[8] Most editions of Wikipedia ban the use of full automatic translation. However, recent application of Lsjbot for automatic translation has led to a huge increase in articles in Swedish along with Cebuano and Waray, languages of the Philippines; if automatization on this scale were to continue, it will have major implications for our understanding of Wikipedia as a site of human interactions.

[9] The basic information about Wikipedia and its current organization and biasses has been gleaned from https://en.wikipedia.org/wiki/Wikipedia; supplemented by Van Dijck (2012) and Rogers (2013). For additional information on editorial practices, see http://contropedia.net/.

Differential memorability (Rigney 2016) is thus at play in what has been celebrated as a global medium. Filtering affects not only the uptake of topics in the different languages, but also the dissemination of particular topics across language areas. To date,[10] there are articles in 34 languages on Alan Kurdi, for example, the Syrian boy whose drowned body became the subject of an iconic photograph in the last months of 2015 (the fact that these articles were produced very soon after the boy's death indicates that Wikipedia is both a "hot" medium of "communicative memory" as well as a medium of long-term "cultural memory"; Assmann 1999). It is interesting to note, moreover, that a majority of these articles (21) were produced in non-European languages (including not just Arabic and Kurdish, but also Malay, Bahasa Indonesia, and two variants of Punjabi), delineating a mnemo-region centered outside of Europe. In contrast: of the 60 articles generated by the massacre of Katyn more than 20 are in Eastern European languages, presumably reflecting an historical interest within the former Soviet bloc in this particular event. Less predictably, however, there are also articles on Katyn in multiple Asian languages from Bengali and Persian to Korean, Chinese, and Vietnamese. It would be extremely interesting to map in more detail such patterns of hetero-remembrance, showing who is interested in whom and for how long (the latter indicated in the number of ongoing edits and page visits[11]). Initial findings suggest there are regional clusters based on geographical proximity and already "scripted linkages" as a result of long-standing historical ties. But they also indicate the occurrence of multiple "unscripted linkages" to geographically distant parts of the world that are not yet part of a familiar narrative, but which may nevertheless resonate by way of analogy with local memories to form de-territorialized mnemo-regions. In the ongoing production of new articles, articulation is thus facilitated, on the one hand, by the affordances of the medium and, on the other hand, by the commitment of local editors to pay attention to particular topics, often using existing articles in another language which they translate and, where relevant, adapt. Conversely, local events can be brought to the attention of an international public by producing articles in English, hence creating

[10] The data relating to Wikipedia presented here was first collected in 2014 and updated in 2019.

[11] Average page visits can be calculated through Wikipedia's own tool: https://en.wikipedia.org/wiki/Wikipedia:Pageview_statistics.

the conditions for them to be linked by way of analogy to resonant events elsewhere.

The distribution of articles dealing with the 1972 Bloody Sunday killing of demonstrators in Northern Ireland can be used to illustrate these patterns of convergence and divergence (Ackermans and Rigney 2015).[12] By 2019, this atrocity had generated acts of hetero-remembrance in 45 languages. These included the languages of the Celtic fringe (Welsh, Galego, Basque, Breton, Irish), an interest that is not surprising in view of "scripted linkages" between these minority groups. But it also included a significant number of articles in Middle Eastern and Asian languages, some of which were direct translations from English, others original articles or hybrids combining translation with new text. A detailed analysis of 20 of these 45 articles, which I conducted with Hannah Ackermans, showed a considerable overlap between the content to which people in different linguistic communities were exposed. The most obvious point of convergence, besides the basic information about the transformation of a demonstration into a site of killing in Derry in 1972, is the tendency for articles to present a list of the victims, including their age (and, depending on the length of the article, some biographical information). While the number and nature of the images varies quite a bit across the articles (6 articles have none), there is also a notable convergence on three images in particular, all of which have to do with the dead demonstrators: a memorial naming the victims (20 articles), a mural showing their faces (15 articles), an installation of fourteen crosses representing each of the victims (14 articles). Alongside these points of convergence, there is also evidence of indigenization, to recall Ramazani, where receiving cultures appropriate and transform content in particular ways.[13]

[12] I am happy to acknowledge here the invaluable contribution of Hannah Ackermans to this section on Bloody Sunday on Wikipedia; in 2014 she collected and compared the different Wikipedia entries then available, providing me with the basic data used here (which I then updated in 2019); see Ackermans and Rigney (2015). At a later stage, Thomas Smits also helped regarding the images used in Wikipedia and how to track them.

[13] Pentzold et al. (2017) illustrate how Wikipedia's neutrality is continuously under pressure from divergent interests and value systems. They point out with reference to bull-fighting that variation between articles can also be linked to ideological positions, with the Spanish article downplaying controversy, the English one drawing more attention to debates about cruelty, and the French one pointing out that bull-fighting is not an exclusively Spanish phenomenon.

To begin with, there is considerable variation in the length of the different articles, ranging from 6000 words in English, to 2879 in French, 2775 in Spanish, 1647 in German, 841 in Turkish, 500 in Malay, and fewer than 200 in Norwegian. This variation indicates that the salience of the topic for particular groups is not necessarily a function of geographical distance or proximity. Apparently, other historical factors make certain groups more receptive than others to particular topics. In addition, a variation in the emotive distance to the events depicted can be noted. One might expect that the geographically closer one gets to the controversy the greater would be the degree of controversy. While it is true that the English-language site indicates a high level of activity around certain issues, the end result, in keeping with Wikipedia's governance model, is that the English-language article is neutral in tone and technical in its register, with evaluative and emotive terms avoided. In contrast (and in keeping with Rogers' findings on Srbrenica), the Dutch and Swedish articles are more emotive in their use of words like "bloodbath" (bloedbad [Dutch]), and "massacre" (massaker [Swedish]), and more insistent in their emphasis on the peaceful character of the demonstration, hence amplifying the unlawfulness of the shooting. This more judgmental engagement with the past indicates the existence of an affective geography that does not neatly coincide with physical distance.

One of the regular features of Wikipedia articles on historical events is the inclusion of links to related art works or works of popular culture. Exemplifying Jenkins' claim about the internet being at the heart of transmedial convergence, the platform works as a relay station between different media and practices of cultural memory (the remediation on Wikipedia of images of the Bloody Sunday memorial and other forms of memorialization already illustrates this basic principle). Comparison between the various articles on Bloody Sunday shows a strong continuity in the links made to the arts, with reference to Greengrass' movie and to the internationally known U2 song *Sunday, Bloody Sunday* occurring in almost all the articles. Their presence reinforces the point made earlier about the role of the arts as mediators between different mnemonic communities since they are presumed to provide a self-evident point of reference and a point of interest for Wikipedia users. Interestingly, several articles also include links to the works of local artists and their acts of hetero-remembrance, with the French and Swedish articles, for example, including invocations of Bloody Sunday on the part of local rap artists. Reflecting a similar "domestication" of the topic, the German article ends with a reference to the fact that the Joseph Beuys prize was awarded in 2010 to a group of artists from Derry. This combination of transnational and local links is replicated in the list of sources where the BBC is regularly found side by side with journalism in the local language of the article.

Another key affordance of the Wikipedia format is the section "categories" which allows the article to articulate its topic to analogous events in the memory culture of the target language. The use of this feature differs quite widely across the articles which each use different categories to link Bloody Sunday to wider historical developments and to frame it as one sort of event rather than another. The Italian article, for example, categorizes the event under the label "state terrorism" and when one clicks through to this category, one finds the Irish Bloody Sunday aligned with the Italian "Anni di piombo" and terrorism in Italy. In the French case, in contrast, we find the category "massacres in the United Kingdom" which then brings us by two degrees of separation to the massacre of Algerian demonstrators in Paris in 1961. Similarly, the German article brings us through the category of "massacre" to Srebrenica and Amritsar. In a comparable way, but now using the "disambiguation" feature of the Wikipedia template, the Turkish article leads us back to other Bloody Sundays, including the 1969 one in Beyazit Square in Istanbul. Finally, in the English article, the disambiguation feature of the basic search term "Bloody Sunday" (1972) shows that this memory site, so central to Irish memory culture, was just one of a series of Bloody Sundays that took place across the world in the last century (see also Rigney 2016).

In these multiple ways, Wikipedia's technological affordances and textual template have intensified the work of articulation that, as I showed earlier, has long been a part of modern memory culture. It generates new and sometimes unexpected links between hitherto dispersed events and lays the basis for new acts of affiliation based on analogy and the comparison between histories. Clearly there is huge potential here for the postnational expansion of memory.

This of course begs the question if and how users realize this potential for articulation. Do people pursue links because they are open to engaging with new stories? Does the Wikipedia interface in itself generate new forms of engagement? Or does it, conversely, merely constitute a huge archive whose activation as "working memory" (Assmann 1999) will depend on the contingencies of reception and of resonance with the narratives already familiar to users? In contrast to the examples given earlier in this essay, Wikipedia does not itself take the form of an experiential narrative capable of generating interest in its own right. This means that understanding its impact in connecting up different mnemonic communities will entail studying the interplay between offline and online memory practices, and between encyclopedic and imaginative forms of engagement with the past as these continue to shape our subjectivities.

Conclusion

I have highlighted the role played by the media, both traditional and digital ones, in "articulating" memories. My examples have shown how mediation establishes virtual connections between "strangers" in such a way as to create virtual mnemo-regions outside traditional frameworks tied to particular territories. They indicate the existence of an emergent mnemotopography which is much more complex than "flat world" models of memory; more complex, not in the least because it demonstrates that memory does not travel evenly across the globe. Even in the case of a multilingual and globally extended platform like Wikipedia, with an extraordinary range and reach, information is channeled selectively, adapted, and indigenized. This is not necessarily a bad thing, however, as long as people continue to actively seek out their blind spots, as citizens and as memory scholars. It is true that mediation often serves to reinforce inherited identities and, in the age of social media, to create echo chambers of the like-minded. However, it is equally true that telling stories, be this in encyclopedic or in experiential form, and inviting subjects to engage actively with them, also helps bring new social relations into being. The ongoing work of articulation, as I've described it here, creates new possibilities for people to relate to those whom they have hitherto considered "strangers", or not considered at all.

In the third decade of the twenty-first century, we need to take this particular role of mediation fully into account in our studies of memory; not just because of its inherent interest, but also because it opens up new perspectives on the capacity of memory cultures to regenerate and to "reinvent" themselves. Changing stories is vitally important in a world where we are increasingly being called upon to learn to live with strangers, migrate into "other pasts" (Huyssen 2003), and take responsibility for the implications of our actions at the other ends of the world. More detailed research is needed into both what enables and what hinders mnemonic articulation as a critical counterpart to the assumption that social frameworks and memories have been fixed once and for all.

References

Ackermans, Hannah, and Ann Rigney. 2015. *Wikipedia as a Medium of Cultural Remembrance*. Working paper posted at http://contropedia.net/.

Amelina, Anna, Thomas Faist, Nina Glick Schiller, and Devrimsel D. Nergiz, eds. 2012. *Beyond Methodological Nationalism*. New York: Routledge.

Anderson, Benedict. 1991 [1983]. *Imagined Communities: Reflections on the Origins and Spread of Nationalism*. London: Verso.

Anderson, Ben, and Paul Harrison, eds. 2010. *Taking-Place: Non-Representational Theories and Geography*. Farnham: Ashgate.

Appadurai, Arjun. 2003. Disjuncture and Difference in the Global Cultural Economy. In *Theorizing Diaspora*, ed. Jana Evans Braziel and Anita Mannur, 25–48. Oxford: Blackwell.

Appiah, Kwame Anthony. 2006. *Cosmopolitanism: Ethics in a World of Strangers*. New York: Norton.

Apter, Emily. 2006. *The Translation Zone: A New Comparative Literature*. Princeton, NJ: Princeton University Press.

Armstrong, Elizabeth, and Suzanna Crage. 2006. Movements and Memory: The Making of the Stonewall Myth. *American Sociological Review* 71: 724–51.

Assmann, Aleida. 1999. *Erinnerungsräume: Formen und Wandlungen des kulturellen Gedächtnisses*. Munich: C.H. Beck Verlag.

Assmann, Aleida, and Sebastian Conrad, eds. 2010. *Memory in a Global Age: Discourses, Practices and Trajectories*. London: Palgrave Macmillan.

Austin, J.L. 1962. *How to Do Things with Words*. Cambridge, MA: Harvard University Press.

Bond, Lucy, Stef Craps, and Pieter Vermeulen, eds. 2017. *Memory Unbound: Tracing the Dynamics of Memory Studies*. Oxford: Berghahn.

Bond, Lucy, and Jessica Rapson, eds. 2014. *The Transcultural Turn: Interrogating Memory Between and Beyond Borders*. Berlin: De Gruyter.

Brubaker, Rogers, and Frederick Cooper. 2000. Beyond 'Identity'. *Theory and Society* 29, no. 1: 1–47.

Crownshaw, Rick, ed. 2014. *Transcultural Memory*. London: Routledge.

Damrosch, David. 2003. *What is World Literature*. Princeton, NJ: Princeton University Press.

De Cesari, Chiara, and Ann Rigney, eds. 2014. *Transnational Memory: Circulation, Articulation, Scales*. Berlin: De Gruyter.

Erll, Astrid. 2011. Travelling Memory. *Parallax* 17, no. 4: 4–18.

———. 2017. Travelling Memory in European Film: Towards a Morphology of Mnemonic Relationality. *Image [&] Narrative: Online Magazine of Visual Narrative* 18, no. 1: 5–19.

———. 2018. Homer: A Relational Mnemohistory. *Memory Studies* 11, no. 3: 274–86.

Ferron, Michela, and Paolo Massa. 2014. Beyond the Encyclopedia: Collective Memories in Wikipedia. *Memory Studies* 7, no.1: 22–45.

Garde-Hansen, Joanne, Andrew Hoskins, and Anna Reading, eds. 2009. *Save As... Digital Memories*. Basingstoke: Palgrave Macmillan.

Grossberg, Lawrence. 1986. On Postmodernism and Articulation: An Interview with Stuart Hall. *Journal of Communication Inquiry* 10, no. 2: 45–60.

Hobsbawm, Eric, and Terence Ranger, eds. 1983. *The Invention of Tradition*. Cambridge: Cambridge University Press.

Hoskins, Andrew. 2011. Media, Memory, Metaphor: Remembering and the Connective Turn. *Parallax* 17, no. 4: 19–31.

Huyssen, Andreas. 2003. Diaspora and Nation: Migration into Other Pasts. *New German Critique* 88: 147–64.

Jenkins, Henry. 2006. *Convergence Culture: Where Old and New Media Collide*. New York: New York University Press.

Kirshenblatt-Gimblett, Barbara, and Jeffrey Shandler, eds. 2012. *Anne Frank Unbound: Media, Imagination, Memory*. Bloomington, IN: Indiana University Press.

Landsberg, Alison. 2004. *Prosthetic Memory: The Transformation of American Remembrance in the Age of Mass Culture*. New York: Columbia University Press.

Leerssen, Joep, ed. 2018. *Encyclopedia of Romantic Nationalism in Europe*. 2 vols. Amsterdam: Amsterdam University Press.

Lévi-Strauss, Claude. 1974 [1958]. *Anthropologie structurale*. Paris: Plon.

Levy, Daniel, and Nathan Sznaider. 2002. Memory Unbound: The Holocaust and the Formation of Cosmopolitan Memory. *European Journal of Social Theory* 5, no. 1: 87–106.

Lowenthal, David. 1985. *The Past is a Foreign Country*. Cambridge: Cambridge University Press.

Nora, Pierre, ed. 1997 [1984–1992]. *Les lieux de mémoire*. 3 vols. Paris: Gallimard.

Peiker, Piret. 2015. History, Politics and Myth: Lydia Koidula's Novella Juudit, or the Last Maroons of Jamaica. In *Novels, Histories, Novel Nations: Historical Fiction and Cultural Memory in Finland and Estonia*, ed. Linda Kaljundi, Eneken Laanes and Ilona Pikkanen, 322–28. Helsinki: Finnish Literary Society.

Pentzold, Christian. 2009. Fixing the Floating Gap: The Online Encyclopaedia Wikipedia as a Global Memory Place. *Memory Studies* 2, no. 2: 255–72.

Pentzold, Christian, et al. 2017. Digging Wikipedia: The Online Encyclopedia as a Digital Cultural Heritage Gateway and Site. *ACM Journal on Computing and Cultural Heritage* 10, no. 1. http://jocch.acm.org/.

Plate, Liedeke, and Anneke Smelik, eds. 2009. *Technologies of Memory in the Arts*. Basingstoke: Palgrave Macmillan.

Ramazani, Jahan. 2009. *A Transnational Poetics*. Chicago: Chicago University Press.

Reading, Anna. 2016. *Gender and Memory in the Globital Age*. Basingstoke: Palgrave Macmillan.

Renan, Ernest. 1947–1961. Qu'est-ce qu'une nation? [1882]. In *Oeuvres complètes d'Ernest Renan*, ed. Henriëtte Psichari, 886–907. Paris: Calmann-Lévy.

Rigney, Ann. 2009. Fiction as a Mediator in National Remembrance. In *Narrating the Nation: The Representation of National Narratives in Different Genres*, ed. Stefan Berger, Linas Eriksonas and Andrew Mycock, 79–96. Oxford: Berghahn.

———. 2012a. *The Afterlives of Walter Scott: Memory on the Move*. Oxford: Oxford University Press.

———. 2012b. Transforming Memory and the European Project. *New Literary History* 43, no. 4: 607–28.

———. 2016. Differential Memorability and Transnational Activism: Bloody Sunday, 1887–2016. *Australian Humanities Review* 59: 77–95.

———. 2021. Remaking Memory and the Agency of the Aesthetic. *Memory Studies* 14, no. 1: 10–23.

Rogers, Richard. 2013. *Digital Methods*. Cambridge, MA: MIT Press.

Rothberg, Michael. 2009. *Multidirectional Memory: Remembering the Holocaust in the Age of Decolonization*. Stanford, CA: Stanford University Press.

———. 2014. Multidirectional Memory in Migratory Settings: The Case of Post-Holocaust Germany. In *Transnational Memory: Circulation, Articulation, Scales*, ed. Chiara De Cesari and Ann Rigney, 123–45. Berlin: De Gruyter.

Rosenzweig, Roy. 2006. Can History be Open Source? *Wikipedia* and the Future of the Past. *Journal of American History* (June): 117–46.

Said, Edward W. 1983. *The World, The Text, and the Critic*. Cambridge, MA: Harvard University Press.

Steininger, Rolf. 2003. *South Tyrol: A Minority Conflict of the Twentieth Century*. Brunswick, NJ: Transaction.

Thaler, Elmar, and Norbert Mumelter, eds. 2011. *24. April 1921, Der Bozner Blutsonntag*. Zirl: Edition Südtiroler Zeitgeschichte.

Van Dijck, José. 2007. *Mediated Memories in the Digital Age*. Stanford, CA: Stanford University Press.

———. 2012. *The Culture of Connectivity: A Critical History of Social Media*. Oxford: Oxford University Press.

Venuti, Lawrence. 2013. *Translation Changes Everything: Theory and Practice*. London: Routledge.

Warner, Michael. 2002. Publics and Counterpublics. *Public Culture* 14, no. 1: 49–90.

Remembering the Violence of (De) colonization in Southern Africa: From Witnessing to Activist Genealogies in Literature and Film

Ksenia Robbe

This chapter engages with the questions of the existence, the characteristics and the dynamics of regional memory practices in Southern Africa, the region stretching from the Republic of South Africa in the south to the borders of the Democratic Republic of Congo and Tanzania in the north. Its focus is on the memories of the decolonization processes that have taken place throughout the region since the 1960s, as mediated by literary and film representations. While there is no shortage of celebratory commemorations of decolonization and regime change, and these can be regarded as a region-wide practice, this reading enquires into more ambiguous and critical remembrance practices that question accepted truths, reveal silenced experiences, and elaborate alternative subjectivities.

K. Robbe (✉)
University of Groningen, Groningen, The Netherlands
e-mail: k.robbe@rug.nl

S. Lewis et al. (eds.), *Regions of Memory*, Palgrave Macmillan Memory Studies, https://doi.org/10.1007/978-3-030-93705-8_8

Decolonization in these works is, then, conceived as a process continuing at present and extending into the future in a variety of ways. At the same time, remembering these periods and rethinking the ways in which the transformations have been framed involve engaging with the nature and continuities of colonial practices.

My discussion zooms into cultural productions from South Africa, Zimbabwe, and Namibia, created over the past two decades, considering them as representative though certainly not exhaustive of the practices and forms of mediating memory in the region. While the novels written in the late 1990s have received considerable critical attention, this reading introduces a comparative perspective on them and juxtaposes these early post-transitional remembrances with the practices of the younger generation that started publishing their works in the late 2010s. It thus charts and analyzes the dynamics of cultural memory across the selected Southern African countries—from the predominance of trauma-related witnessing practices to the development of post-memorial practices in the works of younger writers and film-makers who seek to imagine their (grand)parents' experiences of (de)colonization in order to articulate their own responses to coloniality. Gender appears to be an important factor in these representations. As all of them suggest alternatives to state-supported, official remembrance of decolonization, the memories "from the margins" which they perform involve the perspectives of/on women and those of/on men who do not conform to ideas of hegemonic masculinity.

The part of the Southern African region constituted by the Namibia—South Africa—Zimbabwe triangle represents a heterogeneity of colonial and postcolonial historical experience. The three countries experienced forms of British, Dutch, German, and apartheid colonialism that were, like the decolonization processes that followed, deeply entangled across the region. These entanglements include South Africa's colonial rule of Namibia (1915–1989) and the apartheid state-organized suppression of the anti-colonial struggle in Zimbabwe and Namibia (1960–1980s) as well as, on the other hand, collaborations between national liberation movements and postcolonial political elites. The transition processes involved many similarities and were interdependent: in all three cases, a constitutional framework for a controlled transition was implemented, and in Namibia and South Africa, the transitions involved the change from apartheid to liberal democratic rule (Saunders 2001, 7). Since the 1990s, the countries have been closely interconnected through migration, particularly to South Africa, through cultural interactions facilitated by the common language (English) and the movement of cultural producers and productions (shared audiences and venues of production and distribution). These political and cultural entanglements form a major context and infrastructure for the mediated memories of decolonization to be discussed.

In Search for a Region of Memory: Official vs. Alternative Memoryscapes

The question of whether Southern Africa can be considered a *region of memory* opens a new terrain, especially if one looks beyond the official, state-supported practices of memory and commemoration. The latter became a subject of historical, cultural, and anthropological studies in all three countries (Coetzee and Nuttall 1998; Werbner 1998a, b; Coombes 2003; Marschall 2010; Melber 2003; Becker 2011) and some readings have drawn comparisons between the politics of public history and memory across the region (Melber 2003; Kössler 2010a; Becker 2011). The politics of "elite memorialism" (Werbner 1998a, 8) that silences voices not aligned with the state was first identified and critiqued in the context of Zimbabwe (Werbner 1998b, 71–102). Not only public commemorations and other state-sponsored memorial practices, but also historiography during the 1990s turned into a propaganda mouthpiece with critical voices being excluded and the earlier discourse of "nationalist history" having morphed into one of "patriotic history" (Ranger 2004). A similar trajectory has been traced within the public memory in Namibia, involving "the military image of the liberation struggle, the focus of the victorious and now ruling party and its claim to rule in perpetuity" (Kössler 2010a, 41). The ideologies underpinned by these exclusionary memory practices have legitimated the use of violence against political opponents, to a much greater extent and over a longer period in Zimbabwe, but also in Namibia when a secessionist movement in Caprivi was violently repressed (ibid., 39). In South Africa, the multiculturalist discourse of the "rainbow nation" during the 1990s has often, despite its performed inclusivity, sidelined the experiences of marginalized groups—(black) women, the poor, and the working class—as, for instance, research on the production of testimonies during the work of the Truth and Reconciliation Commission has shown (McEwan 2003; Ross 2003; Krog et al. 2009). Since the 2000s, with the assumed closure of dealing with the apartheid past, the official discourse has shifted towards celebration of the liberation struggle posited at the center of national consciousness, while "the stories of everyday life [have been] subsumed by the triumphalism of struggle history" (Baines 2007, 181).

Considering these similarities, we can trace a memoryscape constituted by nationalist narratives of struggle and martyrdom that foreground the experiences and subjectivities of present-day political elites and their claims to power. However, this memoryscape does not constitute a *region of memory* in the sense proposed by other contributors to this volume, as it does not challenge the imaginaries of a nation: in celebrating the heroism of those who were interacting and collaborating across the national borders during the colonial and apartheid periods, this regime of memory commemorates their struggles only as national actors and frames their sacrifice as committed for the sake of their nation. In contrast, alternative memories of decolonization, even if their framework of reference is mainly national, depict the complexity of subjectivities and memories of those individuals and collectives who interacted across the national and cultural borders. An important factor that distinguishes these latter memory practices, with implications for transnational identification, is their questioning of the continuities constructed in official histories and memories.

The temporal mechanism behind the official memory practices across the region, as has been observed in individual contexts, is the forging of seamless continuity between the earlier and later resistance to colonialism, from the nineteenth century to the present. In Zimbabwe, this has taken the most obvious form through the fashioning of colonial and postcolonial history as a succession of three *Chimurengas*. Chimurenga, the Shona for "revolutionary struggle", was the name of the Shona and Ndebele revolt against the British colonial administration in 1896–1897. The guerilla war that was fought between 1966 and 1979 and resulted in independence was imagined as the Second Chimurenga. What is more contentious, however, is the dubbing of the expropriations of land from white farmers in 2000 as the "Third Chimurenga" by the Zimbabwean government. Furthermore, the recent military coup which displaced Robert Mugabe from presidency involved voices calling it the "Fourth Chimurenga" (Chigumadzi 2018, 14). These acts indicate the compulsion within the society to group varied historical struggles under an unambiguous umbrella of anti-colonial resistance. In Namibia since the 1980s, historians associated with the liberation movement and with the postcolonial state have attempted to construct links between the early twentieth-century resistance and the later struggle of the SWAPO against apartheid (Kössler 2010a, 42–43). Such links are misplaced since different areas of the country have experienced varied colonial invasions which they resisted on their own; the multiethnic nationhood there is only a recent construction.

In South Africa, tendencies of interlinking icons of anti-colonial wars and anti-apartheid struggle can be observed in post-2000 practices of memorialization. An example of this is the Duncan Village Memorial commemorating the 1985 massacre of anti-apartheid protesters while using the symbolism of "traditional" culture by placing a Xhosa warrior at the center of the composition—a misplaced traditionalism that resulted in the community's lack of identification with the monument (Marschall 2012, 196). Moreover, the forging of transhistorical continuities in official memory is often linked to discourses of "unity in diversity", particularly in South Africa. Together, these discursive techniques mask the politics of privileging dominant ethnic groups and bracket out questions of social justice. By creating a multiculturalist façade, this politics of memory works to attract foreign investment and tourism through a growing heritage industry (Witz et al. 2001; McGregor and Schumaker 2006, 655, 657; Becker 2011, 536–38).

Beyond this set of hegemonic discourses and practices, the landscapes of memory in all three countries display multiple contestations and alternatives. These include community protesting against monuments or creating non-intended practices around them (Marschall 2012; Minkley and Mnyaka 2015); developing new curatorial strategies in community museums (Rassool 2006); as well as commemorative rituals and performances such as local history re-enactments and apparel production (Biwa 2010; Kössler 2010b). My reading, however, will focus on literature and film as examples of media that rely on transnational networks and allow for faster exchange across national borders, through distribution and (online) publication mechanisms as well as the movement and collaboration of cultural producers across the region. The potential for regional frameworks of reading Southern African literatures has been explored in the past two decades (Driver 2000; Driver and Samuelson 2004; Weiss 2004; Mwikisa et al. 2010), including in studies of imaginings of the region beyond national borders in literary texts (Primorac 2011; Jackson 2018). None of these studies, however, has focused on questions of memory or considered regional similarities between the ways of dealing with experiences of (post)colonial violence.

This chapter begins such comparison by exploring regional scapes of remembrance beyond the official narratives of unambiguous celebration and enquires into the dynamics of these scapes over time. My reading suggests that cultural productions of the late 2010s reveal a shift in the modalities of remembering the violence of (de)colonization. Hence, the discussion consists of two parts corresponding to this periodization. The first part, engaging with the dominant modality of witnessing in the literature and film of the late 1990s and early 2000s, compares three novels—Kaleni Hiyalwa's *Meekulu's Children* (Namibia, 2000), Yvonne Vera's *Stone Virgins* (Zimbabwe, 2002) and Zoë Wicomb's *David's Story* (South Africa, 2000). I consider them representative of critical memory during this period: they focus both on the experience of liberation wars and anti-apartheid resistance "from below"—involving violence against suspected spies and traitors, as well as against women and ethnic minorities, within the movements and military organizations—and also on everyday violence in public and domestic contexts.[1] Narrative and visual renderings of these experiences, which are insufficiently acknowledged in public memory and often unaccounted for by perpetrators, interrogate the imagined breaks between the colonial and postcolonial or apartheid and post-apartheid regimes.[2]

The second part turns to recent productions by representatives of a younger generation: the poetry volume *Collective Amnesia* by Koleka Putuma (South Africa, 2017), the non-fiction work *These Bones Will Rise Again* by Panashe Chigumadzi (Zimbabwe, 2018) and the film *Unseen* by

[1] An example of a region-wide dialogue through such productions of memory is the series *Landscapes of Memory* (1999) which included documentaries from Namibia, Zimbabwe, Mozambique and South Africa produced by local directors and reflecting specific contexts while being interconnected through themes of trauma, memory and reconciliation (c.f. Mhando and Tomaselli 2009).

[2] The establishment of independent governments and (formal) democracies did not result in the termination of large-scale violence and armed conflict. In Zimbabwe, which achieved independence in 1980, the ZANU-PF confronted resistance by the rival party and responded with massacring around 20,000 of "dissidents" and civilians in Matabeleland North between 1982 and 1987 (operation "Gukurahundi"). In South Africa, the transition period of 1990–1994 saw an escalation of violence resulting in the death toll estimated between 5000 and 14,000 cases which were hardly discussed by the Truth and Reconciliation Commission (1995–1998). The more recent occurrence of state-supported violence against striking miners at Marikana in 2012 demonstrates continuities between the practices of the apartheid and post-apartheid state.

Perivi Katjavivi (Namibia, 2016). While the earlier productions largely rely on tropes of trauma, memories of the new generation who did not experience the time of the "Struggle", involve confrontations with traumas in the form of postmemory. The practices of postmemory in these works generally align with Marianne Hirsch's (2008) theorization and are comparable to second- and third-generation remembrance elsewhere, but involve a stronger activist element and a decolonial consciousness.

Just like the trauma narratives of the preceding decade, this emergent mode invokes breaks and discontinuities in processes of relating to the past. But compared to the earlier writing, their imaginations are more invested in the restorative work of linking to the past in alternative ways. I am referring to such imaginations as *activist genealogies*. The term "genealogy" relates to the Foucauldian method of studying "histories of the present", i.e., of tracing the uncertain processes through which the past becomes the present. However, in contrast to the use of this method to "disturb what was previously thought immobile" and "show the heterogeneity of what was imagined as consistent with itself" (Foucault 1991, 82), the works by the young Southern African authors do not only reveal the colonial roots of contemporary social norms, but also recall moments of empowerment within the colonial past that may suggest how similar subjection to power can be dealt with in the present. This affirmative approach that retraces *decolonial* subjectivities in the history and memory of one's community (usually starting with one's family) is what distinguishes these works. Since the novels from the earlier period—particularly by Vera and Wicomb—have received much critical attention, my reading concentrates on comparing their modes of remembering rather than providing detailed analysis. In contrast, the texts and film analyzed in the second part are very recent and still await critical readings. My consideration of them is, therefore, more detailed though also comparative.

Confronting the Violence of the Struggle: Witnessing and Questioning Reconciliation during the 1990 and 2000s

The novels by Hiyalwa, Vera and Wicomb are among the most innovative and socially critical texts of the turn of the century and the surrounding decades in Namibia, Zimbabwe and South Africa. They are interconnected both thematically and aesthetically: all engage the pitfalls of the politics of reconciliation in the respective countries and draw upon the modes of witnessing and testifying. All three foreground the experiences of violence during the struggle for decolonization that have been silenced and erased from public memory. These counter-memories are narrated and focalized by marginalized figures: young rural women in Hiyalwa's and Vera's novels, and a woman freedom fighter and a man who does not conform to hegemonic masculinity in Wicomb's novel. The texts can be considered narratives of trauma. The use of this mode, and the foregrounding of haunting and the impossibility of fully comprehending and assimilating the past, is political in all three works: they interrupt the practices of "forgetting" this violence within the official memories of decolonization. The novels, thus, converge in questioning the mainstream discourses of reconciliation, but they also, vary in their alternative imaginations of communal and national healing.

These texts co-construct a memoryscape of dealing with the violence suffered by civilians and members of liberation movements across the region. I read this as a regional memory constituted through trauma narratives of gendered suffering, in which representations of women's violated bodies and their similarly disrupted but resilient psyche stand for the condition of postcolonial, post-conflict societies.

Hiyalwa's novel narrates the life of a small village on the border with Angola between 1976 and 1989, through the perspective of a young girl named Ketja who had lost her both parents in an attack by the government army (representing "the boers"); her two younger siblings managed to escape and were transported to military camps in Zambia. Ketja is brought up by her maternal grandmother Meekulu, from whom she learns the virtues of dignified survival and caring for the community. Throughout the narrative, the girl is struggling with the trauma of witnessing the murder of her family, the loss of siblings, and the destruction of the community by the colonial government forces as well as the liberation army, whose soldiers make regular raids into the village and recruit children into the Struggle. Both groups of militants are depicted as bringing devastation into common people's lives: "The colonial army and the freedom fighters [...] had turned our homes into a battle field and we had become victims of the two forces" (54). Yet one of the most poignant points is when Meekulu asserts that for her all "soldiers from the forest" are "her children" (56). Moreover, her statement—"[e]very child in this land is my child. A child of my neighbour is my child. We all feel connected to each other. We are one people." (57)—implicitly extends to the government soldiers (though to Ketja's question if this "people" also includes "the boers", Meekulu responds with "a quiet long look" (ibid.)). This inclusive vision of the "one people" at the time of extreme violence lays the ground for a desired reconciliation that acknowledges the everyday suffering as well as resilience of all Namibians, beyond liberation army fighters.

Vera's novel responds to a somewhat similar situation, though focusing on a more brutal episode, through a narrative that clearly identifies the silenced practices of mass violations in 1980s Zimbabwe. At the same time, it reflects on the entangled origins of this violence, which reveal overlaps between "victims" and "perpetrators." Most clearly this idea is embodied in the figure of Sibaso, who fought the liberation war on the side of the ZAPU and is persecuted by the government forces conducting the massacre. Looking for a place to hide in a village, he kills the protagonist of the preceding narrative, Thenjiwe, and rapes and mutilates her sister Nonceba. The latter survives to bear witness to what her family and community had suffered at the hands of both the "dissidents" and the government forces. Represented through Nonceba's traumatic narrative recalling Sibaso's violation of her body and psyche, the account of his perpetratorship involves the most intrusive and haunting actions and aftermaths. At some points, however, the narrative provides an insight into the logic of this inexplicable violence by foregrounding Sibaso's trauma of the "bush war" and of being persecuted (107), and his ideals of liberation being compromised by the new elites (89). Dorothy Driver and Meg Samuelson read this "violent neurosis" as an example of (postcolonial) melancholia displacing self-destruction upon women and annihilating natural cycles of time associated with them (2004, 186–88). The narrative contrasts this portrayal of melancholia with tentative mourning represented by the figure of Cephas. A former lover of Thenjiwe, he finds Nonceba at a hospital, takes her to his home in Bulawayo and builds a caring relationship with her which involves their shared mourning of her sister. While Sibaso and Cephas' ways of dealing with the past and present of violence are contrasted, the narrative also includes several intersections between the two figures, even making the reader suspect in the scenes of Sibaso's appearance that the killer and rapist might be Cephas, and later forging symbolic connections between these characters which parallel the interconnectedness of melancholia and mourning (ibid., 188–89).

A similar treatment of victimhood and perpetratorship as a space of ambiguity pertains to Wicomb's *David's Story*. Here too, the fighters of the ANC military wing (MK)—those who in post-apartheid South Africa would be expected to be honored as heroes—appear to be the ultimate and paradoxical victims of intimidation and harassment perpetrated by various political (and criminalized) factions within and outside the organization. The story of David's participation in anti-apartheid struggle, however, turns out to be impossible to convey, despite his compulsive desire to have it written down (for which he engages a woman "scribe"). The scribe's discourse as a witness to his psychological struggles ultimately reflects the impossibility of truth-telling. David's recollections for the intended book, she soon realizes, expose multiple silences which he explains by the requirements of secrecy or by the "falsities" of memory (194), but which appear to the readers as signs of his trauma. His narration itself is full of "deleting and rewriting misremembered event[s]" (141). This compulsive re-telling of the past involves references not only to his own suffering and his complex position within the movement as a colored person who, the narrative implies, must have supported "dissidents" within the MK and was tortured for this (195–96; Driver 2000, 236). As a major site of trauma, the narrative invokes the figure of Dulcie—(re)constructed by the writer from snippets of David's talk about a woman comrade—who is subjected to torture by the mysterious people, probably other MK fighters, appearing almost as ghosts in David's narrative (185). In a parallel to the male figures of Vera's novel, Dulcie's torturers reveal an "ambiguity [...], something that makes them both friend and foe as they tend to the cracks and wounds carefully inflicted on parts of the body that will be clothed" (179). This ambiguity extends to David himself who, as the narrative hints, might be complicit in Dulcie's suffering, even if only as a witness (Meyer 2012, 356).

In all three novels, as these cursory readings demonstrate, narratives of trauma posit critiques of reconciliation narratives in Southern Africa. The traumas of decolonization suffered by former combatants are (dis)placed onto women and their bodies, which are violated but also demonstrate possibilities of survival, resilience and recuperation. In their representations of the latter, the novels draw upon female cultural icons. While these images have been engaged by nationalist imaginations too, the novels' narratives re-appropriate them and explore the silenced potentialities of these symbols.

In *Meekulu's Children*, the trope of motherhood is realized through the first-person testimonial narrative of Ketja written as a coming-of-age story, at the end of which she becomes head of the family including her siblings and other children-survivors. The dynamics of the plot—Ketja burying her grandmother and reflecting on her own future role—stresses the motif of appropriation of a gendered national symbol on a community level and matrilineal inheritance. In her embrace of all "children of this land" (56), "Meekulu co-opts a symbol of Namibia the Motherland and claims it as herself, adopting a model of female power which is both maternal and stakes a claim over contested land" (Kornberg 2018, 254). In a similar vein, *The Stone Virgins* involves references to Nehanda, the royal ancestor of the Shona, whose spirit enters the bodies of women-mediums. During the First Chimurenga, she guided the militant Charwe who spearheaded resistance against the British (and was subsequently hanged by them). Vera also explores this figure in her earlier novel published in 1993, *Nehanda*. *The Stone Virgins* includes more indirect references to the prophecies of the spirit's resurrection by alluding to the saying attributed to Charwe/Nehanda ("my bones will rise"): "The bones are rising. Rising." (Vera 2002, 59). Just like the title's reference to the virgins who used to be buried with kings that critiques the sacrifice of women's bodies and subjectivities within Zimbabwe's official memory, citing the words and imagery related to Nehanda signifies a "refus[al] to use women in the way they are used in a nationalist project" (Driver and Samuelson 2004, 195). *David's Story* includes a sub-plot tracing the early twentieth-century history of the Griqua, a Southern African group of multiracial colored people, and their chief Le Fleur, who seems to have been David's great-grandfather. The focus of this exploration is the Griqua's nascent nationalism, which reproduces racist ideologies of Afrikaner and English colonialists with an anti-colonial tint, and particularly the ways this ideology subsumes women's creativity that had been central to the advance of the nation. The narrative parallels these recollected episodes with the gender politics of the anti-apartheid movement.

The narratives of embodied memory—focalizing violence through women's perspectives (Ketja, Nonceba, Dulcie)—perform resistance to nationalist sacrifices of women's bodies. This traumatic memory relies on invocations of witnessing and the mode of testimony, but these narratives avoid essentialist claims to representation; the past becomes accessible only as a trace and through traumatic displacement. Ketja's memory includes several descriptions of her dreams that are populated by figures of the walking dead, mediating her loss and angst, and visions of her siblings participating in the war (their experience of exile never being explicitly conveyed) (52, 72). Nonceba's and Thenjiwe's experiences are narrated as a series of flashbacks, with no clear storyline, focusing on sensations, resonant words and objects. David's narrative is overtly structured through displacement—of his own traumas and the memory of Dulcie invoked only as a trace in his writing about Sara Baartman[3] and as "a scream somehow echoing through [his] story" (134).

The novels, however, part ways in their imaginations of healing or working through trauma; these differences are contingent on the politics of reconciliation in each of the countries. They are most clearly reflected in the narratives' endings. The last scene of Hiyalwa's novel portrays Ketja and Meekulu's other adopted children carrying the grandmother's body to be buried near her house, and being obstructed by government soldiers. Ketja speaks out against them, but the confrontation is ultimately resolved by the arrival of a UN security vehicle and, simultaneously, a group of mourners who help to deliver the corpse and the children to their old/new home (Hiyalwa 2000, 117–19). This hopeful ending signifies the birth of a new community through Ketja adopting the role of a symbolic mother for *all* children. In contrast to the mainstream narrative of the Namibian transition which proposed to forget the past of colonialism and war (in conjunction with blanket amnesty), the novel posits the memory of the civilian victims—Ketja's parents and grandmother killed by the colonial army—at the center of the nation's regeneration. "From now on", Ketja reflects, "I will have the task—a very big responsibility—to show the children the resting place of their parents and relatives and where many children of Elombe are buried." (113). In Vera's novel, Nonceba's gradual recovery, facilitated by Cephas' care (also in his symbolic capacity as a historian reconstructing the precolonial image of Bulawayo) can be read as a metaphor for the nation's recuperation. Yet her return to the normality of life is only part of what needs to happen. In Cephas' thought, "What he sees in Nonceba is only what is recoverable … It would be too much to ask for her to be entire. It would be impossible." (Vera 2002, 158). This, again, should be read against the background of the state's superficial politics of mourning which recognizes the "excesses" of post-liberation violence only to silence their remembrance. Wicomb's judgement of post-apartheid memory politics is even more somber. David's and Dulcie's stories keep haunting the writer, defying representation and recovery; both

[3] Sara Baartman was a Khoikhoi woman who was exhibited at human zoos in Britain and France during the early nineteenth century and examined by anatomists as a "link" between animals and humans.

end in suicide and the writer's resolution to "wash [her] hands of this story" (Wicomb 2000, 213). This defiance involves a sharp critique of post-apartheid practices of remembrance and jurisprudence, represented by the granting of amnesty in exchange for truth-telling during the TRC, which was generally framed as a therapeutic process necessary for the new nation to come into being.

Thus, the memoryscape of representing the 1980 and early 1990s transition processes in Southern African fiction is defined by deconstruction of the victim/perpetrator dichotomies and of heroic tropes in official discourse. Instead of the binary visions, it foregrounds multifaceted everyday experiences of war and conflict. It focalized these events through the subjectivities of women, some of whom are combatants, i.e., it also interrogates the constructed boundary between civilians and soldiers, war heroes and survivors. While all examined representations rely on and elaborate tropes of trauma, the novels also stress the agency of (female) survivors. This is achieved through tracing (imaginative) matrilineal genealogies to women ancestors who are represented as symbols of perseverance and care for victims and perpetrators (*Meekulu's Children*); as reclaimed icons of anti-colonial leadership whose image has been abused by a nationalist ideology (*The Stone Virgins*); and as powerful, emancipated women whose role in liberation struggle has been obliterated even within a liberal narrative of transition (*David's Story*).

Although these novels (and other representations from the 1990s and 2000s elaborating similar plots and tropes) have drawn public attention to the gendered violence of decolonization and transition, this memory has been only marginally incorporated into mainstream discourses and practices. The official remembrance of the period today shows little difference to that of the earlier decades, even despite the recent displacement of Robert Mugabe from his 27-year rule of Zimbabwe and the stepping down of Jacob Zuma from presidency in South Africa. Over the past decade, remembering colonialism and decolonization has become a major theme in popular mobilizations and activist practices which challenge the rule of the parties that grew out of liberation movements.[4] In most cases,

[4] In South Africa, the most prominent example has been the #RhodesMustFall movement against the colonial monuments. In Namibia, there exists activism for the remembrance of the Herero and Nama genocide of 1904–1908; the German government officially recognized this event as genocide in May 2021. In Zimbabwe, decolonial memory practices involve drawing on traditional spirituality and ideas of justice in addressing colonial and postcolonial violence orchestrated by the state (Morreira 2016, 57–88).

however, approaches that identify colonial practices in post-independence and post-apartheid politics and draw on indigenous ideas of justice have not been embraced by larger publics or (sufficiently) supported by international organizations. These approaches, nevertheless, are being actively developed within contemporary artistic practices including literary writing and film, to which my discussion now turns to consider the continuities as well as departures of the younger writers and film-makers from the earlier representations of memory. In the times of post-transitional disappointment regarding the possibilities of influencing postcolonial memory politics, where do they locate hope and agency?

Voices of the "Born Frees": Postmemory and Genealogies of (De)coloniality

In the following discussion of works by a writer, a poet, and a film-maker, produced over the second half of the 2010s, I use "born free"—the Southern African idiom referring, optimistically, to the generation born after decolonization or the end of apartheid—as an umbrella term (the three authors were born in the 1980s and early 1990s). I regard their work as representative of the emerging aesthetics and politics of the 2010s and of the younger generation's voices within it. While most of their productions are focused on the authors' respective countries (those where they or their parents are from), South Africa as the regional powerhouse constitutes an important context for their life and work. Panashe Chigumadzi was born in Zimbabwe and emigrated with her parents to South Africa in 1994. She grew up and still lives there whilst keeping strong connections to Zimbabwe. Perivi Katjavivi, a son of a prominent Namibian politician and leader of the SWAPO in exile and an English-born, naturalized Namibian who played a key role in the Namibian publishing post-1989, was born in Oxford and grew up in Windhoek during the 1990s. He received an MA and is doing a PhD in South Africa. South Africa's regional "pull" in these cases is reflective of the economic and educational patterns of (trans-)migration, whereby many people spend their time between the neighboring countries. Also, much of the publishing and film industry is concentrated in South Africa, which explains the strong presence of the authors in this country. The choice of the authors for the current discussion is determined by the "visibility" of their works nationally and transnationally as well as the similarity of decolonial memory practices that I regard as defining the shape of an emerging regional memory.

Whereas the earlier regional memory was centered on witnessing and testimony, the experience of the younger generation leaves little space for such practices. Their works, however, foreground trauma as a condition shared by the younger Southern Africans and one which has deeper roots than their own or even their parents' generational experience. Kim Wale, Pumla Gobodo-Madikizela, and Jeffrey Praeger reflect on this societal condition in South Africa as "post-conflict haunting", which involves "not simply a memory of past violence, but a lived, everyday experience of continued suffering and exclusion" (2020, 2). The works of literature and film discussed below convey this sensibility which is increasingly invoked in Southern African contexts by the younger generation who voice disappointment with the lack of political will among the post-transitional elites to tackle the burning issues of social inequality. In South Africa, where reconciliation was an institutionalized state-supported project, and where, thus, the expectations of its outcomes were particularly high, the sense of betrayal and of being haunted by the colonial/apartheid past has been bitter. This sensibility started taking larger collective shape in the wake of the Marikana massacre in 2012, when 34 striking mineworkers were shot down by the police—a turning point in post-apartheid society that for many revealed the uncanny continuities of the present with the apartheid past. The growing sense of the continuing coloniality post-2012 gave shape to the massive student protests in 2015–2016 under the slogans of #RhodesMustFall and #FeesMustFall (Naicker 2016). An integral component of these protests and related activism were the practices of visual art, photography, and poetry, mediating the imaginaries of recalling resistance to apartheid in the past, in the context of the continuing coloniality of the present. These works of art and poetry involve a strong mnemonic dimension as they "recall the aesthetics of anti-colonial expression and infuse it with more contemporary decolonial inquiry into the ways of knowing the past" (Robbe 2016, 453). Photographs taken by students during the protests, as Kylie Thomas observes, constitute the "acts of decolonial world-making" through the "new iconography of resistance" that "draws on, references and re-animates the past and yet breaks away from the social documentary forms of representation that characterized the struggle against apartheid" (2018, 99). At the same time, in Namibia, memory activists representing the Herero and Nama communities have been challenging the post-independence politics of remembering centered on the national myth of anti-apartheid struggle that obscures the longer histories of violence and anti-colonial resistance (Harmick and Duschinski 2018).

In the present examination of the texts and films that appeared at the time of social movements and memory activism against coloniality, I propose that the new imaginaries and iconographies that involve decolonial practices of (post)memory—what I call here "activist genealogies"—are being articulated in literature and film on a regional scale. Just like photography and art, these texts relate to the earlier modalities of witnessing in literature through a dialectics of suffering and hope, performed through representing trauma and reclaiming the marginalized agency of female cultural icons or family figures. The difference of the more recent productions lies not only in their shifting of trauma from the mode of witnessing to the mode of postmemory,[5] but also in their understanding of current traumas as shaped by the longer practices of coloniality—from the first encounters with colonialism to the present day. Along the lines of decolonial critique, my reading refers to "coloniality" as an epistemological characteristic of modernity that posits Eurocentric knowledge and its subject as universal and continues doing so after the end of colonial rule.[6]

In terms of form, works by the younger authors gravitate towards hybrid genres, enhancing documentary writing with elements of fiction and vice versa: Chigumadzi's book combines features of memoir, journalistic writing and poetic prose; Putuma's Spoken Word poetry includes narrative elements (a bullet point list in "No Easter Sunday for Queers" and "On Black Solidarity"; an essay structure in "Oh Dear God, Please! Not Another Rape Poem"); Katjavivi's film merges conventions of documentary (the genre that has been prevalent in Namibian productions such as Richard Pakleppa's) and feature film. Related to the hybridity of form and of generational memory is the focus of these productions on everyday life in postcolonial societies, where boundaries between ordinary experience

[5] Marianne Hirsch's term that "describes the relationship that the generation after those who witnessed cultural or collective trauma bears to the experiences of those who came before, experiences that they 'remember' only by means of the stories, images, and behaviors among which they grew up" (Hirsch 2008, 106).

[6] "Decolonial critique" here refers to the practices of delinking from the colonial matrix of power in attempting to establish socio-cultural relations of a different kind. This critical approach draws on Franz Fanon's analysis of continuing colonialism and the pitfalls of decolonisation and develops it with regard to present-day conditions while imagining ways of overcoming this regime (Quijano 2000; Maldonado-Torres 2007).

and politics are less distinct than during the earlier times of anti-colonial struggle or military conflict.

Koleka Putuma's debut collection *Collective Amnesia* (2017) is one of the most outright and vivid expressions of black South Africans', and particularly women's, experiences, struggling with present-day manifestations of coloniality. The title refers to the push of the mainstream post-apartheid discourse to simply leave the colonial past behind, thus rendering invisible the continuity of colonial power structures within the society. This involves more than a manipulation of the black majority by the whites who continue holding on to power; this discourse also includes the patriarchal bias within the black communities, which does not allow for expressions of women's agency, as well as the older generations' silencing of pain and suffering as a way of coping. Thus, speaking out against the colonial structures of the present, the poems reflect on black people's complicity in their reproduction. The shaping of alternative ways of remembering is a major agenda of Putuma's work as a poet and theater-maker; in her own words, "[s]he is particularly interested in how narrative making can be a catalyst of healing, resistance and excavating memory".[7]

A major theme in the collection is the trauma effects of the younger generation resulting from living in an African society which is thoroughly "white" in its treatment of experiences and perspectives that diverge from the Eurocentric mold. The opening poem "Black Joy" represents a fond childhood memory of spending time at grandparents' home together with siblings and cousins and sharing simple food and entertainments. It is interrupted by a reflection on the perspective of a white observer that infiltrates the minds of those who are remembering:

> But
> isn't it funny?
> That when they ask about black childhood,
> all they are interested in is our pain,
> as if the joy-parts were accidental. (13)

The scenes of everyday life in this poem do not romanticize poverty, but they present an alternative to the stereotypical images of black suffering that sharply contrast the apartheid past and the post-apartheid present and render the experience of the "born frees" as a clean slate. A similar

[7] https://cocoputuma.wordpress.com/about/.

struggle is staged in "Hand-Me-Downs", which recalls the practice of passing on one's clothes and things to younger siblings. The chain of sharing extends to the entire community and places the speaking subject within a lineage (the inheritance of objects metaphorically connects with the process of remembering—"Inherited memory" is the title of the cycle): "I also come from a lineage of borrowed and borrowing./The neighbour's sugar was an open jar without a debt collector." (15) The metaphor culminates in a dehumanized image of black people who are treated like second-hand things by the culture of coloniality: "Cheap./ Disposable./ Damaged./Labour." (17) Just like the previous poem that shows both the pain and joy of living in a township, this one depicts the two sides of the culture of second-hand things: the thirst for the new and the expensive along with the experience of humanity that such sharing entailed and that seems to be almost forgotten (18–19).

The theme of traumatization as a result of postcolonial amnesia is further developed in "Water"; here the current state of distress acquires more historical depth by being related to the traumas of slavery and colonialism. The poem opens with the stereotype of black people being afraid of the sea and swimming. The pain of being mocked and thinking of oneself through the images of the "*Baywatch*-like" is related to the embodied memory of slaves transported and swallowed by sea (96). The state of trauma is, then, displaced from humans onto the sea which is being "re-traumatised" every time black bodies touch its surface—traditionally, during public holidays (97). The traumatic memory of enslavement is only enhanced by the present in which the relevance of this memory is denied, and black people "are asked to dine with the oppressors/ and serve them forgiveness" (99). The speaking subject's response is: "How, / when the only ingredients I have are grief and rage?" (ibid.). The reference to December 16, the Day of Reconciliation which was introduced in 1995 to combine memorable dates of the Afrikaner community (victory over the Zulus in 1864) and of black South Africans (a peaceful anti-apartheid protest and the founding of Umkhonto we Sizwe in 1961) stresses the colonial aspects of the negotiated settlement of the 1990s and of the public memory it attempted to construct (is it possible for black South Africans to celebrate their defeat and colonial conquest?). The speaking subject's rejection of the narratives of transition is rendered with sarcasm in "1994: A Love Poem", in which she asks to be loved "the way that white people look at and love Mandela" (101). This iconoclastic rendering denies the functioning of "1994" as a historical watershed and re-focuses our attention on the persisting coloniality of discourse and memory.

It is, however, not only the white hegemony but also the older generation's ways of dealing with pain and experience of oppression by simply silencing them that Putuma's poems seek to debunk. Speaking about her generation in first person plural, she laments the fact that "we are taught/ that mourning is the opposite of strength" (56). "Graduation" renders a similar theme within the more personal scope of a mother-daughter relationship while at the same time addressing the two female characters as generalizable "you" (35). A later stanza introduces the generational difference between the women, revealing the daughter's discomfort with the mother's silencing of loss (37).

The recognition of the validity of the older generation's approaches implies a caring yet resistant attitude and the claiming of autonomy. The dissociation from her mother only leads the subject to figure an alternative, queer womanhood, as she does in "Reincarnation", through a reflection of a militant image of her grandmother (20). Such connections appear vital, as rendered in "Lifeline" which is a list of names of black women writers, scholars and activists from South Africa and elsewhere that constitute for the speaking subject "a gospel shut in [her] bones" (85). This imaginary lineage—a recurring image in the collection—represents an alternative memory of women's place in history and suggests a starting point for decolonial narratives. In her recent reading of Putuma's collection, Elleke Boehmer compares its style and mode of address to the 1970s poetry of Mongane Serote which was "radical, hard-hitting, and raw, openly powered by Black Consciousness values of race pride and anger at the apartheid system" (2018, 180). This aesthetic and political connection to an earlier generation of anti-apartheid (male) poets co-constructs the openly queer and decolonial genealogical imagination in Putuma's work.

In ways similar to the performance of women's lineage in Putuma's poems, Chigumadzi's book engages with the memory of Nehanda. This homage to Yvonne Vera's writing is indicated in the phrase from her novel *Nehanda*, "history is created in the mouth", placed in the epigraph. The author's exploration begins with her personal memory of a photograph (the only copy of which has been lost) of her grandmother Lilian Chigumadzi after she passed away in October 2017. A month later, Zimbabwe was experiencing a historical moment when a coup displaced Robert Mugabe and the vice president of the ruling party took his place—the few weeks that were celebrated by many as the "Fourth Chimurenga".

Chigumadzi's view of this change is skeptical as it has brought to power yet another military leader with a background and approach to politics similar to Mugabe's. In reflecting on the transformation, her project is to proceed from the memories of her grandmother and other women some of whom, like Nehanda's mediums, have been elevated to the status of Mothers of the Nation while others—like Grace Mugabe against whose succession as a president the coup had occurred—have been reduced to "Evil Stepmothers, birthing all that is bad in our society" (109). She decides to "lower [her] eyes from the heights of Big Men who have created a history that doesn't know little people, let alone little women, except as cannon fodder" (19) and re-imagine this history through memories of women, reflecting on the symbolical and physical violence inflicted on them during the 1980s and the recent transition.

This project of undoing the violence begins with the author's memory of Lilian's photograph as a sixteen-year-old girl, taken at the Panokromatic Studio in Mutare in 1956. The remembered photograph becomes a springboard for the author's imagination, who regrets never being able to speak to her grandmother about her life. The narrative takes Panashe on a journey to her grandmother's home village during which she visits other relatives, friends, and Nehanda's current mediums. She recalls the photograph as having fascinated her with the image of a young woman who was "[a]lone, unburdened, unattached, only Belonging to Herself" (30). Later she finds out that Lilian was a "born-rukesheni" (42), i.e., born and grown up in a township, which implied an experience of cosmopolitan modernity during the 1940–1950s. Associating this personal history with the 1956 song by Dorothy Masuka, which tells a story of a girl who wore trousers but was accepted by her lover nevertheless (53–54), she reflects on the necessity of remembering the experiences of women "who refused their places in time" (55). By (re)creating her grandmother's story as part of a cultural memory of Zimbabwe's modernity, the text seeks to liberate the memory of "women's resistance in rural areas, cities, mines and farms before, during and beyond the Chimurenga" that have been "foreclosed by the singularity of Mbuya Nehanda's 'frozen image'" (92). The idiom of a frozen image is borrowed from Vera's reflection on Nehanda's only photograph taken before her medium was executed during the First Chimurenga (Hunter 1998, 77). First fixed by a colonial camera, it has then been appropriated by anti-colonial nationalists and the postcolonial state, which turned it into a national symbol and put Nehanda's present-day mediums under strict control (Chigumadzi 2018, 76–86).

Against this background of family and national memory, the narrator begins contemplating ways out of the cycles of violence, including the 2017 "military-assisted transition" (94). In Zimbabweans' cultural memory, she observes, the postcolonial nation's birth is associated with mass violence and deaths, including the sacrifice of Nehanda (ibid.). Thus, violence seems to have become the core of the national and individual identity—a realization that makes the narrator ask:

> Is our fixation with Nehanda's death a cathartic confrontation of our colonial trauma? Or is it an acceptance of violence as being always necessary in our making and remaking? Perhaps we continue to rehearse Nehanda's execution because we have not found the way to resolve the traumas of our violent past and present? (95)

Against this time of traumatic repetition and of the opposition's "presentist tradition" that does not offer a vision beyond Mugabe's displacement, she reflects on ideas of timelessness and transcendence as a foundation for reimagining history as accessible to everyone. Its democratic potential is conceived as beginning with a women's lineage echoing Putuma's poem: "Chimurenga transcends time and space, belonging to our ancestors, our grandmothers, our mothers, ourselves, and to our daughters and granddaughters. Instead of reclaiming Chimurenga for our people, our opposition conceded it to Big Men" (99).

This democratic imagination transcends both the moment of decolonization and the recent transition as turning points. Instead of seeing "Zimbabwe's future .... [as] a matter of the old dying and the new being born" (129), it proposes a vision of cyclical time involving "cumulative and intergenerational traumas and triumphs" (130). It also involves a transcultural regional imagination as opposed to exclusive ideologies of "Shona-ness" or even "Zimbabwean-ness". Through her own migrant subjectivity, she reflects on the similarity of Zimbabwe's and South Africa's postcolonial discourses which had figured the moment of decolonization as "zero time" and imagined the new generation as being "born free" (63). An opening up of this continuity of colonial time occurs when the narrator finds out, through her grandfather's slip of the tongue, that her paternal ancestors had migrated, in the eighteenth century, from the present-day KwaZulu-Natal in South Africa to the eastern part of Zimbabwe (44). Such discoveries of regional fluidity are intertwined with imaginations of timelessness, and together they form spatio-temporal coordinates for new regional identities.

Perivi Katjavivi's film *Unseen*, in a similar vein, displaces the vision of Namibia's independence as a moment of the nation's birth. Following the three protagonists stranded in the urban spaces of Windhoek, it reveals that decolonization had left them in a limbo from which there is no escape. The stories of Anu, Marcus and Sara form separate plotlines which are interconnected through the sense of isolation and disconnectedness. Anu is a local rap musician who feels alienated from his former friends and the community of the "location" in which he grew up; he also realizes that there is no way for him to fit into the global music industry, even though he raps in English. His girlfriend tries to find an escape by sleeping with rich men, though every time she is dumped, she returns to Anu. Marcus, a black US actor, becomes one of these men, spending a brief period in Windhoek on invitation to play Mandume, the king of the Kwanyama who was either killed or committed suicide during the South African army's attack on his kingdom in 1917. The film shows Marcus's struggle to relate to the material of a different culture and time. At the end, the news that the Namibian director had failed to secure funding for the production confirms his sense of the futility of travelling to the small African country and of the alienation he experiences there.

The disjunction between the global and the local in these two stories is translated into the third plotline that features a depressed young woman, Sara, attempting to commit suicide in her apartment. Together, the three stories elaborate the metaphor of the postcolonial "unseen"—the local which, though speaking a global language, remains virtually invisible and ungraspable to both vernacular and transnational audiences, and the silence that surrounds the idiosyncrasies of "minor" histories. This condition of invisibility without end also appears as a state of being stranded in time. In his reading of recent (transnational) South African writing, Andrew van der Vlies describes this temporality as "no-time"—"a present that did not feature in the past's privileged narrative of time's future unfolding" (2017, 5–6). This temporality disrupts the dominant post-apartheid/ postcolonial imaginations of a new beginning; the past future itself appears to be colonized. The film conveys this idea most clearly in Sara's dialogue with a man who attempts to make her speak by asking: "What are you running from? Doesn't it ever get lonely?", to which she responds: "One can race around, race towards something, but there are limits, at the heart of the world, behind or beyond".

An alternative to this is articulated in Anu's monologue in one of the last scenes when, lying on bed with his girlfriend, he starts contemplating a different time to counter their disappointment in the present: "How far back do you think our childhood goes? Fifty years, a hundred, a thousand? I feel like it goes farther back than the memory of any race. Our dreams come from, like, deep below, you know. From a deep remembering of, like, forgotten ancient ancestry or something." Accompanied by meditative music, his narrative invokes the Pyramids of Giza as the greatest human achievement belonging to Africans and imagines his girlfriend as Isis reborn. This is the only moment in film when the camera stops capturing the characters or their environment and focuses on photographs: the viewers see a slide show interspersing the pyramids with early twentieth-century photographs of the enslaved and colonized people of Namibia. Thus, Anu's imagination, evoking "deep time" and a Pan-African reach, addresses the colonial past and its implications in the present. Drawing on the legend of Osiris and Isis with its empowering motif of rebirth, this imagination is hopeful, despite the persisting colonial trauma and bitterness: "So we've got a raw deal, ha?", asks the girlfriend. "The rawest. But it's cool, though", replies Anu.

Conclusion

While existing research has indicated similarities between official memory practices in post-1990 Southern Africa and particularly Zimbabwe, Namibia and South Africa, this chapter has asked whether we can trace an alternative regional memoryscape that encompasses reflections on the violence and suffering of decolonization during the 1980 and 1990s (with references to the colonial periods and the continuing practices of coloniality). The readings have, thus, focused on literature and film that involve critiques of mainstream memorialization and invoke alternative constellations of past, present and future. Focusing on different generations of authors enables a consideration of the historical dynamics of such alternative remembrances, and an engagement of very recent works that have not yet received much scholarly attention.

The region of memory that this chapter traces across the three Southern African countries is constituted through the commonality of mnemonic frameworks through which the shared historical experiences of colonial and apartheid violence, decolonization, and "post-conflict haunting" are mediated. My reading has traced the dynamics of such a memoryscape from the practices of witnessing and testifying to the violence of decolonization towards the recent articulations of memory that create activist genealogies of tackling coloniality across the periods of resisting colonialism, anti-apartheid struggle, and the contemporary critique of post-transitional/post-independence politics. The examined practices of literature and film are part of the larger societal processes of engaging with past violence and resistance: the literature of witnessing was inherently linked to the projects of truth-telling and reconciliation (institutionalized within the South African TRC and represented by smaller-scale projects in other Southern African countries) while the imaginary of activist genealogies is involved, as I pointed out above, in contemporary protest movements. In both cases, works of literature and film create common tropes and critiques of engaging with the past, and due to their "portability" (Rigney 2004), facilitate regional circulation of these mnemonic practices.

The comparative reading of three acclaimed novels published around the year 2000 demonstrates a series of similarities in their representations of anti-colonial struggle culminating during the periods of transitions and in their immediate aftermath. These include the narratives' focus on the everyday experiences of male and particularly female protagonists involving the violence of war and military resistance. These representations show how violence entwines "victims" and "perpetrators" into ambiguous relationships and shared traumas. While traumatic tropes prevail in these novels, they also stress the agency of the (women) protagonists and differ in their imaginations of the possibilities for healing and reconciliation.

The narratives of the younger generation re-engage with imaginations of decolonization and transition, but zoom out of the modes of witnessing to create larger, though no less personal, perspectives. What Boehmer observes regarding Putuma's poetry can be considered as characterizing the *region of memory* in recent Southern African writing and film: these productions "transport [readers] kinesically through and beyond the contested present moment" (2018, 175). In other words, they create frameworks of embodied memory that focus on structural violence and its *longue durée* in the postcolony.

Both the older and newer productions dismantle the imaginations of the end of colonialism and apartheid as moments of nations' rebirth, when the past can be laid to rest and a reconciled people can begin history anew. Focusing on the present-day experiences of injustice, yet locating their roots in the earlier, colonial periods, Putuma, Chigandzi and Katjavivi's works displace narratives of deliverance and (extended) transition. In them, the traumas of decolonization are perceived as traumas of coloniality—of the past relations that reproduce themselves in the present. Based on this vision, they refuse understanding transitions of three decades ago as a radical break. These narratives, thus, posit a perspective *after* the hope for "recovery", but also *against* the impossibility of hope. Their search for hopefulness involves re-imagining the hidden, previously un-narrated stories of everyday life, survival, and community.

The emerging practices of cultural memory analyzed in this chapter do not only communicate the trauma of (de)colonization but seek to understand the sources of the entire society's traumatic relationship to both the colonial past and the past of anti-colonial struggle. While the earlier texts, as my reading has shown, elaborate connections to (female) ancestors beyond official memorialization, these newer works radicalize—in decolonial ways—the idea of *timelessness* that comes with such alternative re-assembling of past and present. Explicitly, as in Chigumadzi's text, or implicitly, they propose conceptions of circular time which defy colonial and Eurocentric linearity in favor of metaphors of permanent movement and regeneration, which also re-imagine the region as a *historically fluid space*. They lend transnational, regional form to what Richard Werbner has referred to as local practices of "popular history in which the past is perceived to be unfinished, festering in the present—[...] the narratives which motivate people to call again and again for a public resolution to their predicament" (1998a, 9).

Acknowledgements I would like to acknowledge the support of the Polish Institute of Advanced Studies in Warsaw and the EURIAS Fellowship Programme of the European Commission (Marie-Sklodowska-Curie Actions—COFUND Programme—FP7) during my research and writing of this chapter.

References

Baines, Gary. 2007. The Politics of Public History in Post-apartheid South Africa in History Making and Present-Day Politics. In *The Meaning of Collective Memory in South Africa*, ed. Hans Erik Stolten, 167–82. Uppsala: Nordiska Afrikainstitutet.

Becker, Heike. 2011. Commemorating Heroes in Windhoek and Eehana: Memory, Culture and Nationalism in Namibia, 1990–2010. *Africa* 81, no. 4: 519–43.

Biwa, Memory. 2010. Stories of the Patchwork Quilt: An Oral History Project of the Nama-German War in Southern Namibia. In *The Long Aftermath of War –*

Reconciliation and Transition in Namibia, ed. André Du Pisani, Reinhart Kössler and William A. Lindeke, 331–70. Freiburg: Arnold Bergstraesser-Institut.

Boehmer, Elleke. 2018. The Transformative Force of the Postcolonial Line: Protest Poetry and the Global Short Story. In *Postcolonial Poetics: 21st-Century Critical Readings*. 173–92. Basingstoke: Palgrave Macmillan.

Chigumadzi, Panashe. 2018. *These Bones Will Rise Again*. London: Indigo.

Coetzee, Carli, and Sarah Nuttall. 1998. Introduction. In *Negotiating the Past: The Making of Memory in South Africa*, ed. Carli Coetzee and Sarah Nuttall, 1–15. Cape Town: Oxford University Press.

Coombes, Annie. 2003. *History After Apartheid: Visual Culture and Public Memory in a Democratic South Africa*. Durham/London: Duke University Press.

Driver, Dorothy. 2000. Afterword. In *David's Story*, ed. Zoë Wicomb, 215–71. New York: The Feminist Press.

Driver, Dorothy, and Meg Samuelson. 2004. History's Intimate Invasions: Yvonne Vera's *The Stone Virgins*. In *The End of Unheard Narratives*, ed. Bettina Weiss, 175–208. Heidelberg: Kalliope paperbacks.

Foucault, Michel. 1991. Nietzsche, Genealogy, History. In *The Foucault Reader*, ed. P. Rabinow, 76–100. New York: Pantheon.

Harmick, Ellie, and Haley Duschinski. 2018. Enduring Injustice: Memory Politics and Namibia's Genocide Reparations Movement. *Memory Studies* 11, no. 4: 437–54.

Hirsch, Marianne. 2008. The Generation of Postmemory. *Poetics Today* 29 , no. 1: 103–28.

Hiyalwa, Kaleni. 2000. *Meekulu's Children*. Windhoek: New Namibia Books.

Hunter, Eva. 1998. 'Shaping the Truth of the Struggle': An Interview with Yvonne Vera. *Current Writing* 10, no. 1: 75–86.

Jackson, Jeanne-Marie. 2018. Reading for the Region in New African Novels: Flight, Form and the Metonymic Ideal. *Research in African Literatures* 49, no. 1: 42–62.

Katjavivi, Perivi. 2016. *Unseen*. Film. https://vimeo.com/150133368.

Kössler, Reinhart. 2010a. Images of History and the Nation: Namibia and Zimbabwe Compared. *South African Historical Journal* 62, no. 1: 29–53.

———. 2010b. Political Intervention and the Image of History: Communal Memory Events in Central and Southern Namibia. In *The Long Aftermath of War – Reconciliation and Transition in Namibia*, ed. André du Pisani, Reinhart Kössler and William A. Lindeke, 371–402. Freiburg: Arnold Bergstraesser-Institut.

Kornberg, Netta. 2018. Power at the Margins: Black Female Agency in Two Namibian Novels. In *Writing Namibia: Literature in Transition*, ed. Sarala Krishnamurthy and Helen Vale, 241–61. Windhoek: University of Namibia Press.

Krog, Antjie, Nosisi Mpolweni and Kopano Ratele. 2009. *There was This Goat: Investigating the Truth Commission Testimony of Notrose Nobomvu Konile*. Pietermaritzburg: University of KwaZulu-Natal Press.

Maldonado-Torres, Nelson. 2007. On the Coloniality of Being: Contributions to the Development of a Concept. *Cultural Studies* 21, no. 2–3: 240–70.

Marschall, Sabine. 2010. *Landscape of Memory: Commemorative Monuments, Memorials and Public Statuary in Post-Apartheid South Africa*. Leiden: Brill.

———. 2012. Memory and Identity in South Africa: Contradictions and Ambiguities in the Process of Post-Apartheid Memorialization. *Visual Anthropology* 25, no. 3: 189–204.

McGregor, JoAnn, and Lyn Schumaker. 2006. Heritage in Southern Africa: Imagining and Marketing Public Culture and History. *Journal of Southern African Studies* 32, no. 4: 649–65.

McEwan, Cheryl. 2003. Building a Postcolonial Archive? Gender, Collective Memory and Citizenship in Post-Apartheid South Africa. *Journal of Southern African Studies* 29, no. 3: 739–57.

Melber, Henning. 2003. 'Namibia, Land of the Brave': Selective Memories on War and Violence Within Nation Building. In *Rethinking Resistance: Revolt and Violence in African History*, ed. Jan Abbink, Mirjam de Bruijn and Klaas van Walraven, 305–27. Leiden: Brill.

Meyer, Michael. 2012. Replaying Trauma with a Difference: Zoë Wicomb's Dialogic Aesthetic. In *Trauma, Memory and Narration in the Contemporary South African Novel*, ed. Ewald Mengel and Michela Borzaga, 349–63. Leiden: Brill.

Minkley, Gary, and Phindezwa Mnyaka. 2015. Seeing Beyond the Official and the Vernacular. In *The Politics of Heritage in Africa: Economies, Histories, and Infrastructures*, ed. Dereck Petersen, Gavua Kodzo and Ciraj Rassool, 50–69. New York: Cambridge University Press.

Mhando, Martin, and Keyan Tomaselli. 2009. Film and Trauma: Africa Speaks to Itself through Truth and Reconciliation Films. *Black Camera* 1, no. 1: 30–50.

Morreira, Shannon. 2016. *Rights After Wrongs: Local Knowledge and Human Rights in Zimbabwe*. Palo Alto: Stanford University Press.

Mwikisa, Peter, Mary Lederer, and Leloba Molema. 2010. Introduction. *Research in African Literatures*, Special Issue on Southern African Literature 41, no. 3: v–x.

Naicker, Camalita. 2016. From Marikana to #feesmustfall: The Praxis of Popular Politics in South Africa. *Urbanization* 1, no. 1: 53–61.

Primorac, Ranka. 2011. Dialogues Across Boundaries in Two Southern African Thrillers. *The Journal of Commonwealth Literature* 46, no. 1: 157–72.

Putuma, Koleka. 2017. *Collective Amnesia*. Cape Town: Uhlanga.

Quijano, Annibal. 2000. Coloniality of Power, Eurocentrism, and Latin America. *Nepantla* 1, no. 3: 533–80.

Ranger, Terrence. 2004. Nationalist Historiography, Patriotic History and the History of the Nation: the Struggle over the Past in Zimbabwe. *Journal of Southern African Studies* 30, no. 2: 215–34.

Rassool, Ciraj. 2006. Community Museums, Memory Politics, and Social Transformation in South Africa: Histories, Possibilities, and Limits. In *Museum Frictions: Public Cultures/Global Transformations*, ed. Ivan Karp, Corinne A. Kratz,Lynn Szwaja, andTomas Ybarra-Frausto, 286–321. Durham: Duke University Press.

Rigney, Ann. 2004. Portable Monuments: Literature, Cultural Memory, and the Case of Jeanie Deans. *Poetics Today* 25, no. 2: 361–96.

Robbe, Ksenia. 2016. Erinnerung als Waffe der Dekolonisierung: Kunst und StudentInnen Bewegung im heutigen Südafrika. Trans. Gerhard Hauck and Reinhart Kössler. *Peripherie* 144 (36): 432–54.

Ross, Fiona. 2003. *Bearing Witness: Women and the Truth and Reconciliation Commission in South Africa*. London: Pluto Press.

Saunders, Christopher. 2001. From Apartheid to Democracy in Namibia and South Africa: Some Comparisons. In *Transition in Southern Africa – Comparative Aspects, Two Lectures*, 6–17. Uppsala: Nordiska Afrikainstitutet.

Thomas, Kylie. 2018. Decolonization if Now: Photography and Student-Social Movements in South Africa. *Visual Studies* 33, no. 1: 98–110.

Van der Vlies, Andrew. 2017. *Present Imperfect: Contemporary South African Writing*. Oxford: Oxford University Press.

Vera, Yvonne. 2002. *The Stone Virgins*. Harare: Weaver Press.

Wale, Kim, Pumla Gobodo-Madikizela, and Jeffrey Prager. 2020. Introduction: Post-Conflict Hauntings. In *Post-Conflict Hauntings: Transforming Memories of Historical Trauma*, ed. Kim Wale, Pumla Gobodo-Madikizela, and Jeffrey Prager, 1–25. Basingstoke: Palgrave Macmillan.

Weiss, Bettina, ed. 2004. *The End of Unheard Narratives*. Heidelberg: Kalliope paperbacks.

Werbner, Richard. 1998a. Beyond Oblivion: Confronting Memory Crisis. In *Memory and The Postcolony: African Anthropology and the Critique of Power*, ed. Richard Werbner, 1–17. London/New York: Zed Books.

———. 1998b. Smoke from the Barrel of a Gun: Postwars of the Dead, Memory and Reinscription in Zimbabwe. In *Memory and The Postcolony: African Anthropology and the Critique of Power*, ed. Richard Werbner, 71–102. London/New York: Zed Books.

Wicomb, Zoë. 2000. *David's Story*. New York: The Feminist Press.

Witz, Leslie, Ciraj Rassool, and Gary Minkley. 2001. Repackaging the Past for South African Tourism. *Daedalus* 130. no. 1: 277–96.

Transoceanic Entanglements: Remembering Forced Labor Migration in M.G. Vassanji's *The In-Between World of Vikram Lall* and Yvonne Adhiambo Owuor's *Dust*

Hanna Teichler

> *MY NAME IS VIKRAM LALL. I have the distinction of having been numbered one of Africa's most corrupt men, a cheat of monstrous and reptilian cunning. To me has been attributed the emptying of a large part of my troubled country's treasury in recent years. I head my country's list of shame. These and other descriptions actually flatter my intelligence, if not my moral sensibility.*
> —Vassanji 2003, prologue

It is through this drastic, yet somewhat ironic and mildly self-congratulatory characterization that the reader of M.G. Vassanji's novel *The In-Between World of Vikram Lall* (2003) makes the acquaintance of its eponymous

H. Teichler (✉)
Institute of English and American Studies, Goethe University, Frankfurt a.M., Germany
e-mail: teichler@em.uni-frankfurt.de

S. Lewis et al. (eds.), *Regions of Memory*, Palgrave Macmillan Memory Studies, https://doi.org/10.1007/978-3-030-93705-8_9

protagonist. This prologue forms the introduction to Vikram's colorful story as he, a descendant of indentured laborers shipped in from India, makes his way through life and searches for his position within the Kenyan social fabric. His rise to questionable fame is told to us in retrospect and from his rather comfortable position in Canadian exile. As he looks back on his life, his past achievements and failures, the reader accompanies him and the Kenyan nation through their various stages of coming-into-being. Vikram's childhood, his formative years at university and his rise to power within the new Kenyan state are set against the backdrop of Kenyan history in flux: Vikram witnesses the demise of British colonialism, the Mau Mau uprising as resistance against the British colonial forces, and Jomo Kenyatta's independent Kenya.

Most significant to him and by extension to (his idea of) Kenyan national identity is the legacy of Indian indentured labor; the remembrance of how his ancestors built the East African Railway. Through a remarkable act of resignification, as will be further outlined in this essay, the traumatic legacy of forced labor migration is transformed into the origin story of the Indian diaspora in Kenya. Thus, *Vikram Lall* testifies to the transnational, transregional and transcultural configuration of Kenya's national identity.

In a similar manner to *Vikram Lall*, Yvonne Adhiambo Owuor's novel *Dust* (2014) pursues an extensive mnemonic agenda to paint a picture of Kenya's multiple (hi)stories. The text sends its main protagonist, Ajany Oganda, onto a painful quest to make sense of her brother's death. Set in Kenya in the early 2000s, *Dust* takes the assassination of Odidi at the hands of corrupt police forces as the starting point of Ajany's search for truth. Her journey to find who and what caused his untimely death becomes her pilgrimage to "find" the independent Kenyan nation. The text skillfully juxtaposes representations and representatives of British colonialism with those of a quite corrupt Kenyan state in which foreign rule may have disappeared, but has been replaced by a new elite which seems unfazed by the overwhelming sense of disappointment in Kenyan independence. It is a story of search and partial retrieval, but also of profound loss, disappointment, and death.

Every generation, the narrative tells us, has lost someone or something to the Kenyan struggle under the colonial yoke, and later for independence. Nyipir, Ajany's father, for example, repetitively references the fate of his siblings who did not return home from the theatres of the Burma

Campaign of World War II. In what historian David Killingray references as a forgotten narrative of the Second World War, African colonial soldiers and auxiliaries played a key role in Great Britain's war in East Asia (2001, 426).[1] These African men who enabled strategic victories alongside European soldiers, and saw colonial forces defeated, later played a central role in African nationalist movements and campaigns for independence, as they were mythologized as unintentional "freedom fighters". Nyipir's departure to search for his long-lost brothers marks the end of novel, on which occasion the text remarks: "*What endures?* Starting again." (Owuor 2014, 362; emphasis in original.)

Of particular analytical interest to the present volume on regions of memory, however, is the way this novel subtly taps into the memory of slavery. On occasion of the loss of her brother, Ajany returns home from Brazil, where she had previously moved to. Brazil holds a particular significance in the context of remembering colonialism. In the course of its duration, more than four million African slaves were shipped to present-day Brazil via transatlantic slave trade routes. The Triangular Trade system not only had a devastating effect on African cultures and peoples, but forcibly propelled Africans into the world.[2] The Atlantic gradually became a memory region that is defined by "Atlantic readings of slavery where the boundary between enslaved and free is unequivocally demarcated and racialized" (Hofmeyr 2012, 722). Consequently, African diasporas became the focus of interest for both scholarship and literary engagements (see e.g. Gilroy 1993; Araujo et al. 2011; Olaniyan and Sweet 2010; Clark Hine 2009; Eckstein 2006). This memory of forced migration to Brazil is not only implied in Owuor's novel, I argue, but also undergoes a process of resignification. *Dust*, which is centrally concerned with exploring post-independence Kenyan national identity, reframes the legacies of slavery as part and parcel of Kenya's mnemonic landscape, but with a twist. Although the text does not provide a clear reason for her journey to Brazil, it is a

[1] Around 600,000 men and women from Britain's African colonies formed the King's African Rifles, Or East African Rifles; during the Second World War, over one million Africans put on uniforms (c.f. Killingray 2001, 427). In late 1942, for example, African soldiers were deployed as frontline troops against the Japanese (c.f. Killingray 2001, 428). As Killingray further outlines, going overseas towards the East was a "startling experience" for most African soldiers (2001, 435). These combatants were later heralded as having fought abroad for freedoms "unknown to them at home" (2001, 426).

[2] The demarcation "African" serves as an umbrella term here, and does suggest conflate the many cultures, languages, ethnicities and memories that comprise the African continent.

highly symbolic move on the part of the narrative: it is Ajany's *choice* to relocate there. By looking outward and following the traces of the Transatlantic Slave Trade, the novel suggests, Kenya looks back at itself (see Mwangi 2009). Instead of reiterating narratives of victimization, the old routes of the Atlantic slave trade become avenues of remembrance but also of possibility. Yet, her stay in Brazil results in a traumatic experience as she ostensibly kills her ex-partner in an act of psychological self-defense—another identity-politically charged twist in the narrative that requires further analysis.

In a similar manner to *Vikram Lall*, Ajany's journey into her own past (and that of her brother) becomes the conduit to represent the multi-layeredness of Kenya's national narrative: the legacies of forced labor migration are present in their historical and traumatic dimensions, yet no longer serve as the primary identity template. Their memories are indeed interwoven into the national fabric as origin stories, as sources to rethink, represent and reinvent Kenyan identity in all its ambiguity. *Vikram Lall* arguably romanticizes the construction of the East African Railway as it becomes the origin story for Kenya's Indian community, and Ajany's move to Brazil might seem like the expression of a privilege held by the new Kenyan elite that the text so succinctly criticizes. Indeed, national identity here is ultimately tied to the legacies of forced labor migration as historical and imaginative forces connecting countries and continents such as Africa, India, Canada, Britain, and Brazil.

Yet, read in juxtaposition, these texts challenge the traditional separation between the Indian and the Atlantic Ocean as separate mnemonic spaces, or regions of memory; the former associated with the practices and legacies of indentured labor, the latter with transatlantic slave trade. The comparative framework that allows for, or more precisely calls for a juxtapositional reading of both novels, is that of English-language literatures across the globe, which understands anglophone literature not as an "imaginary assembly of national literatures" contained within themselves, but as "a discursive field [… that] is capable of encompassing both the literary practice of writers who can no longer be related to one particular 'national literary space' […] but also to the worldwide field of English-language literatures and specific forms of communicative interaction and political conflict engendered by it" (Schulze-Engler 2012, 11–12). Put differently, Owuor's and Vassanji's texts are case studies to (further) theorize comparative frameworks of a global reach. *Vikram Lall* and *Dust* thus exemplify how fictional representations of transregional entanglements operate with what Astrid Erll describes as "conscious modes of relational remembering connecting various cultural memories" (2018, 274) across time, and more importantly for this analysis, across space. As my reading will show, these two texts render visible historical, mnemonic and imaginary connections that are situated at the threshold between the national and the global. In their efforts to imagine a national consciousness after Kenyan independence (*Dust*) and to dwell on the positions of coolie descendants in the Kenyan national fabric (*Vikram Lall*), these novels represent a Kenyan national identity that is radically facing outward, across the oceans. As will be outlined in the following, in order to analyze how remembering chattel labor makes visible multidirectional entanglements cutting across national and cultural frameworks as well as mnemonic regions, I argue, a shift of perspective—from the regional/national to broader lenses such as the transoceanic—is required.

Kenya Unbound

On July 21, 2017, former president Uhuru Kenyatta awarded the status of Kenya's forty-fourth tribe to the Indian community. This gesture of acknowledgment followed a request from this very collective to be formally recognized as part of the Kenyan national fabric. By virtue of this petition, "[t]he Head of State acknowledges that the Kenyan Asians' Contribution to Kenya has its roots at the dawn of the nation" and further invites this group to continue to "participate in the political, economic, cultural and social development of the nation" (Mwere 2017). This political performance is indeed a reference to the long-standing connections between Africa and Asia in general, and between Kenya and India in particular. Before, during and after colonialism, people, goods and stories moved to and fro between the continents, effectively cutting across national, cultural and political boundaries (Desai 2010; Hofmeyr 2012; Allen 2014; Njenga Karugia 2018). Seen in this light, Uhuru Kenyatta's gesture of recognition appears like an arguably belated, but logical political affirmation of such connections.

What is more, awarding this status to the Kenyan Indian community is a remarkable act of memory politics. Although not specifically mentioned, the colonial past and its memory discourses are implicitly present as part and parcel of Kenyan national history but are now reframed as a foundational narrative for a diverse national identity. This act recognizes Asian indentured labor—a system of bonded labor that was officially instituted following the abolition of slavery—as fundamental to Kenya's development as a nation. Indentured laborers were recruited to work on sugar, cotton and tea plantations, and rail construction projects in British colonies in the West Indies, Africa and South East Asia. From 1834 to the end of the First World War, Britain had transported about two million Indian indentured workers to nineteen colonies including Fiji, Mauritius, Ceylon, Trinidad, Guyana, Malaysia, Uganda, Kenya, and South Africa. The indentured workers (known derogatively as "coolies") were meant to receive wages, a small amount of land and, in some cases, a return passage once their contract was fulfilled.

One of the enduring legacies of Asian chattel labor in Africa is that of the East African Railway. Indeed, this project effectively entangled agendas and identities from its inception onwards. The East African Railway was mostly built by Indian and Chinese forced laborers; trains were manufactured in England and Scotland, then shipped over the Atlantic and the Indian Oceans to reach their destinations in Africa. German engineers supervised the undertaking and helped to maintain it. Moreover, this project was infamous for its enormous death toll, brought about by natural forces such as man-eating lions and unbearable working conditions. Yet, the successful completion and expansion of the project rendered it a transnational East African success story, all the more so after the formal end of colonialism. The train lines are still vitally important today, and more Afrasian connections are forged through this project. China has apparently offered to fund the extension of trackage, effectively connecting Kenya, Uganda, Rwanda, and Tanzania (Anyanzwa 2019).

The East African Railway legacy thus is a narrative that is shared across national and regional borders. As Astrid Erll poignantly argues, such shared histories can become "relational objects", providing an opportunity for communities to form connections across national or cultural boundaries (Erll 2018, 275). With regard to contemporary Kenyan memory politics as evidenced by awarding the Indian diaspora tribe status, the legacy of the East African Railway is highly ambiguous. On the one hand, it testifies to colonial intrusion and its impact, but on the other hand, it has also come to represent resilience against the dominance of colonialism, both as a violent force and as a dominating narrative. Thus conceived, the East African Railway not only formed the basis for the emergence of a shared history between African nations (Uganda, Tanzania, Kenya) but also of a common "imagery between Africa and Asia" (Achenbach et al. 2020, 5).[3] On an identity-political level, memories of Asian chattel labor form the basis for reframing this legacy as a source of "Afrindian" collective identities (c.f. Njenga Karugia 2018), effectively producing what Arjun Appadurai succinctly calls "long distance nationalisms" (1996, 22) by creating an Afrasian transregion (c.f. Achenbach et al. 2020) through remembering the construction of the East African Railway. More precisely, such entangled memory discourses facilitate the imagination of a Kenyan national identity that is facing outward, across national boundaries and across the Indian Ocean, spanning a physical and imaginative space that John Njenga Karugia poignantly refers to as the "Afrasian Sea" (2018).

M.J. Vassanji's novel *Vikram Lall* is a fictional testimony to and exploration of these ambiguous Afrasian connections. Interestingly enough, Vassanji is frequently labeled as a Canadian author. Whilst Canada certainly plays a recurrent role is Vassanji's fictions, his oeuvre cannot be bound to national containers and national literatures, neither Canadian nor Kenyan. Characteristic of Vassanji's oeuvre is the representation of movement and mobility, sometimes along old colonial routes, but always across demarcations such as Global North and South.

Moreover, Vassanji ostensibly belongs to the group labeled "Indian Ocean writers". In recent years, such fictions have become increasingly more visible. A foundational text with an oceanic perspective certainly is Joseph Conrad's *Heart of Darkness* (1902), a chilling exploration of European colonialism in East and Central Africa with the Thames and the Congo River as watery and imaginative extensions of the Indian and Atlantic Oceans. Amitav Ghosh arguably is the most prominent contemporary Indian Ocean writer, having dedicated large parts of his oeuvre to representing the (post)colonial histories and identities that reside, linger, travel, and come into being in this formative oceanic space. Nobel laureate Abdulrazak Gurnah's novels are frequently regarded as exploring such Afrasian connections as well as renegotiating Africa's relationship with Europe and the former colonial powers. This arguably incomplete list shall

[3] There is an increasing body of scholarship on African diasporas in India that is interested in exploring the legacies, effects and present states of African presence in India. For further reading, see e.g. Dubey 2010; Jayasuriya 2003 and 2010. Such a perspective on Afrasian entanglements across the Indian Ocean scholarship is also reflected in the Frankfurt-based research cluster AFRASO (Africa's Asian Options, http://www.afraso.org) which set out to explore the "complexities of African-Asian interactions" (n.p.).

serve to give an idea of how established a genre Indian Ocean literature has become.

Vassanji's novel *The In-Between World of Vikram Lall* is centrally concerned with migration and mobility on the Indian Ocean and beyond. The frame narrative features Vikram in his Canadian exile, where he was sent after a corruption scandal was discovered, and lets him reminisce about his life with all its achievements and failures. He feels marginalized, trapped in Canada, and yearns to redeem himself in the eyes of his family and his nation. The reader follows this potentially unreliable narrator as he episodically remembers his life, and events that are significant to the development of an independent Kenyan nation are also significant to his personal development.

Vikram Lall is a second-generation descendant of coolie railway workers and son of a traditional Punjabi family. His grandfather came to Kenya as a chattel worker to build the East African Railway. Vikram's parents find themselves sitting somewhat between the cultural stools, trying to maintain what they perceive to be an authentically Punjabi culture while openly sympathizing with British rule and lifestyle. They are framed to represent a middle class family who have made a modest fortune in Kenya, yet who struggle to identify as Kenyan. Vikram, the family's pride and joy, rises through the ranks of independent Kenya, until all comes to an end and he is exiled from Kenya.

British colonialism is present in the novel in all its ambiguity. Vikram's father is the proud owner of a convenience store selling camembert and English cookies, who would sit "with his ears glued to the radio, listening to the BBC's live commentary [of the coronation of Queen Elizabeth] taking place in London at Westminster Abbey" (Vassanji 2003, 71). Vikram confirms this emotional attachment to the colonial center with exclaiming "[h]ow proud we were to be her subjects then, to belong to the mighty empire (ibid.)". This close connection to the Queen and Commonwealth yielded a specific nickname for Kenya's Indian diaspora:

> Here I was, a young Asian graduate in an African country, with neither the prestige of whiteness or Europeanness behind me, nor the influence and numbers of a local tribe to back me, but carrying instead the stigma from a generalized recent memory of an exclusive race of brown "Shylocks" who had collaborated with the colonizers. (Vassanji 2003, 238)

Consider how Vikram in a seemingly natural manner refers to himself as Asian although he was born and raised in Kenya. What is more, the reference to European Jewry and the implied memory of the Holocaust not only serves to underline his marginalized status within the Kenyan nation, but points towards a multidirectional connection of memory discourses that is so characteristic of the novel (c.f. Rothberg 2009). Remembering and making sense of these memories does not unfold within the tight frameworks of the national or cultural. At this point, Vikram sees no reason to identify as Kenyan but rather engages in cross-referencing identities and (traumatic) histories. This episode quoted above is followed by Vikram musing how he came to be part of the Ministry of Transport, first as a comptroller—a job suitable for satisfying the "assumption that as an Indian I was naturally adept in handling large sums of money" (Vassanji 2003, 239). From this point on, Vikram rises through the ranks of the independent Kenya of Jomo Kenyatta, joining the East African Railways.

Central to Vikram's rise within the independent Kenyan nation, and to his own sense of self, is the memory of how his grandfather left Punjab and built the railways. One day Vikram takes his aging grandfather to see the trains one more time:

> In a thick cloud of clinging hot steam in the early morning chill of the Kenya Highlands, the 5607 "injun", as the Sadarji engineer calls his locomotive, pulls off under the watchful eyes of the Kijabe station crew [...]. Beside me stands Dadaji, shaky on his feet, clutching my arm as he looks down at the valley, which we descend at a slow ten miles per hour, his eyes glazed with grim nostalgia. This is the route he toiled sixty-five years ago with fellow Punjabi labourers, here he lost the tip of his pinky finger as the rails were laid down one after the other on the muddy slopes during March rains. (Vassanji 2003, 233)

Here, the East African Railway becomes a legacy to reflect on with nostalgia. The loss of grandfather's pinky finger is an interesting choice of reference to the hardships of Asian indentured labor—the constant threat to health and life is turned into an unpleasant, yet somewhat bearable mnemonic legacy. What seems to be more important to Vassanji's protagonists is the sense of belonging to Kenya, a notion emerging from the almost sugar-coated memory of the East African Railway. Thus conceived, the grandfather's memory of coolie labor is gently transformed into the *raison d'etre et de rester* of Kenya's Afrindians.

Consider how the traumatic and painful dimension of indentured labor is completely absent from the narrative when Vikram himself, now employed by the East African Railway, reflects on what this job means to him and family:

> According to my family, my arrival at the East African Railways was pure destiny—kismet and karma combined, sheer good fortune and a just reward. We had come full circle, from my grandfather laying down the rails at the inception of the railway to myself, assistant auditor and inspector on the line in independent Kenya. No other job could have thrilled me so much. As a boy I had dreamed of speeding on a railway engine from lake to coast, crossing the country back and forth, head and shoulders leaning out proudly to appraise the world flying past before me, like those Sadarji engineers I admired. I would imagine trains travelling west from Nairobi to Lagos and Accra, south to Cape Town, north to Khartoum and Cairo, uniting all Africa. […] The country was mine to explore, on this mysterious metal highway stretching from the coast into the interior, its iron rails reaching to diverse, far-flung and strange places; stories clung to it and ghosts still haunted its path." (Vassanji 2003, 239–40)

Here, the East African Railway becomes the origin story of Kenya's Indian diaspora. In fact, it becomes increasingly hard to maintain the diaspora label—Kenya's Indians were an integral part to building the very structure that enabled mobility across East Africa and has become a foundational narrative to Kenya's "success story". Consider how the narrative creates a straight line, a teleological connection, from colonialism and indentured labor to the present in independent Kenya. Such connections unfold on what Astrid Erll would identify as a vertical axis, effectively entangling memories of colonial Kenya and expectations in independent Kenya (2018). Mnemonic hierarchies—the vertical axis—undergo a reordering. To negotiate, construct and carve out Kenyan identity is no longer tied to the colonizer-colonized paradigm; Kenyanness is no longer primarily measured against Britishness. Instead, trans-African connections and Afrasian entanglements are central here; relations beyond the colonial power dynamics become visible and important. This shift of focus testifies to the fact that there "is no memory prior to relationality"; but memories are the vehicle to render visible such connections (Erll 2018, 279). The quote above represents the "emergence of a shared imagery between Africa and Asia" (Achenbach et al. 2020, 5) and testifies to the "multiple memberships" that both individuals and collectives engage in (Erll 2011, 10; c.f.

Sen 2006). In this sense, this framing of the legacies of indentured labor not only becomes the foundational narrative of an Afrindian or Afrasian identity, but also marks the beginning of a new Kenyan identity, ushering in a sense of independence from the colonial legacy. Thus, in *Vikram Lall*, imaginations of independent Kenyanness stretch well beyond Kenya's own historical and geographical borders.

Another example of how contemporary fiction writers reconfigure received mnemonic narratives and "update" them for the demands of the present is Yvonne Adhiambo Owuor's debut novel *Dust*. Owuor was born and raised in Kenya. She later moved to England to study for an MA. In contrast to M.J. Vassanji, Owuor is considered a Kenyan author and marketed as such. In a similar manner to Vassanji, however, her fictional work is interested in exploring non-canonical relationships and identities forged by transoceanic entanglements.

Dust engages with another traumatic colonial legacy that indisputably left its violent imprint on the world map, and on many cultures: slavery and forced chattel migration across the Atlantic Ocean. The triangular trade encompassed a large portion of the world and was connected through the Atlantic. Slave ships from Britain left ports like London, Liverpool, and Bristol for West Africa, carrying goods such as cloths, guns, and ironware that had been made in Britain. Later, on the West African coast, these goods would be traded for men, women and children who had been captured by slave traders or bought from African chiefs. On the brutal "Middle Passage", enslaved Africans were densely packed onto ships that would carry them to the West Indies (c.f. Araujo 2012; Araujo et al. 2011). Transatlantic slave trade was custom since the 1500s but saw its heyday during the eighteenth century. An estimated 10 to 12 million forced labor migrants were shipped out of Africa alone.

Multidirectional memory cultures have formed around the representation of slavery (c.f. Rothberg 2009). Due to its impact, brutality and longevity, the transatlantic slave trade is the most formative historical legacy in the context of human migrations and in relation to a maritime space. It is a history that is visible today but also heavily contested, for example as a context for the Black Lives Matter movement in the United States. By the same token, "(neo)slave narratives" have become a literary genre in their own right and have garnered an international audience. Caryl Phillips, for example, has come to be known as one of the most prolific and visible authors engaging with the legacies of slavery. Phillips, who was born in St. Kitts, raised and educated in London, and currently resides in the USA, chronicles the triangular movement of the black diaspora in his oeuvre. Appraised by critics and readers, he has been dubbed the "bard of the Atlantic" and a "citizen of the world", for his imaginaries are as much concerned with retelling the stories of the slave trade and the Middle Passage as with pointing out their global reach. The toll of the triangular slave trade was famously set in relation to the Holocaust by American writer and Nobel prize laureate Toni Morrison's 1987 novel *Beloved*: the paratext to what has been called a neo-slave narrative reads "60 Million and more" (Morrison 2007 [1987]). Most recently, Esi Edugyan's novel *Washington Black* (2018) claimed a lot of critical and popular attention as a new example of representing the slave as an avantgarde cosmopolitan. Many more can be listed here but suffice it to state that the outright cruelty taking place on the Atlantic and via the transatlantic trade triangle has been very visible in recent years.

In a similar manner to Vassanji's *Vikram Lall*, the memory and routes of slavery and transatlantic trade are interwoven into the fabric of Owuor's novel and repurposed with regard to their meaning for a contemporary Kenyan identity. The main protagonist's brother, Odidi Oganda, is killed by the Kenyan police on the streets of Nairobi. His death occurs in the night of the 2007 elections and will eventually serve as the token to represent Kenya's many failures and shortcomings as an independent nation. Corruption is rampant, violence is omnipresent and omnipotent, no one seems entirely guilty or innocent. After Odidi's tragic death, Ajany returns home from Brazil where she had spent some time and, as later becomes apparent, thought to have committed a murder. The Oganda family reunites over Odidi's death in the novel's present, yet they are constantly thrown back into their past. To represent the entangled nature of Kenyan memory discourses, Owuor allocates a voice and a history to many members of this "new" state: the reader encounters a colonial officer whose misguided affection for Kenyan treasures leads to the disaster of origin that forms the basis for the Oganda family history. There are tribesmen—Turkana, Luo, Somali, Eritrean—who contribute pertinent perspectives to the Kenyan kaleidoscope that Owuor unfolds. It is a highly ambiguous national identity that Owuor allegorizes through this novel, one that is tied to the legacies of the past and whose present is oftentimes still impacted by the remainders of colonialism and its power imbalances. Yet, this text is particularly powerful because it refuses to remain caught in a perennial struggle of negotiating Kenya's relationship with England and romanticizing independent Kenya. On the contrary, *Dust* succinctly criticizes what has become of Kenya, carving out in detail to which extent a new Kenyan middle class is complicit in the nation's current unfortunate state and dubious affairs.

Ajany's life story holds the most significance for this analysis. As already mentioned above, Ajany returns home from Brazil. This country holds a specific significance in the context of slavery and transatlantic slave trade. A large number of slaves forced to cross the Atlantic Ocean arrived in Brazil; some sources estimate that up to 50 per cent of all human cargo arrived at the shores of this South American country (c.f. Klein and Luna 2010). Moreover, Brazil was the last country in the region to abolish slavery in 1888. The Afro-Brazilian connection was brought about by slave trade and forced labor migration. *Dust* is a poignant example of how literary fiction binds memory and space: it subtly hints at longstanding historical connections and traumatic legacies while repurposing their memories for imagining contemporary independent Kenyanness.

This legacy of Afro-Brazilian connections is referenced within *Dust* but has been turned upside down. It is Ajany's choice to move from the Kenya of the late 1990s and early 2000s to Brazil where she explores her identity as an artist, a woman and an African, and enjoys the nightlife. Note how heterogenous the crowd is that Ajany encounters:

> By way of the window into Nairobi night skies, memory had dragged [Ajany] back to the terrain in Brazil called "Saudade". Its actual name was "Clube Dorival". It was evolved by a few of the assorted casualties of music's broken promises. [...] Portugal-infused Africans, vagabond refugees, wounded immigrants, all-in-betweeners, representatives of the nations' detritus, those who had disappeared into "lost" and the merely curious. It was a place of meeting, and sometimes a bordello. Ajany found it five years ago, on an evening of drizzles, after a dishevelled woman stepped into her homeward path, palm out, and snarled "Menina" in a too-high voice. (Owuor 2014, 138)

Saudade, the Portuguese word for longing that is more precisely understood as *Weltschmerz*, becomes the operative word in this quote. Perhaps it reflects the desire for finding one's history, and, in these troubled waters of histories intersecting, for attaining a sense of arrival, closure and completeness. What both the visitor of this club and the reader encounter, however, is a prolonged state of in-between-ness, and a sense of weariness at the world's ambiguity that can be tormenting at times. Here, Brazil becomes the culmination point of many individual and collective histories: "Dark-brown décor, coffee, wine, voices into music, over music, as music, russet mustiness. A blend of spice and perfume, night jasmine, too. Aphrodisiac melodies" (Owuor 2014, 139). With a simple succession of words, *Dust* implies and simultaneously juxtaposes the memories of forced labor on plantations, and ancient spice trading across the waters, with the narrative present of Ajany, the protagonist and focalizer of this scene. In this vein, *Dust* reverses the trajectory that the trade routes of slavery determined for its human cargo—out of Africa, and into the "New World". Ajany willingly leaves Kenya only to be forced to return on occasion of Odidi's death. Brazil as the former destination for labor forces who were denied their right to be human, to be regarded as human, becomes the space to explore humankind's pleasures and pains, agendas and uncertainties.

The significance of the Brazilian episode in Owuor's novel can be accessed through its relevance for the novel's main theme: Kenyan national identity and its relationship to (colonial) history. Ajany's journey to and experience of Brazil are thus part of Kenyan national identity, referencing its colonial history. Put differently: Brazil is part and parcel of Kenyan identity because they remain forever entangled by the legacies of colonialism. Entangled, however, does not mean rendered immobile, on the contrary: this part of the plot thus reveals more about the text's rendering of a Kenyan self-image as closely connected with the yonder than about the lingering trauma of slavery. It stands as exemplary for how memories and their representations are "produc[ed] [...] across belief systems, spatial imaginaries, and temporal and emotional regimes" in order to suit the demands of a highly situational present, in this case Kenyan national identity (Erll 2018, 276). The possibility as well as the desire to travel the ancient slave routes as members of the rising Kenyan middle class represents the foundational position that the history of slavery holds in this context, but also points towards process of re-signifying this traumatic legacy to shape an African identity that moves beyond the colonizer-colonized dichotomy.

Read in juxtaposition, both novels thus construct a Kenyan national identity that understands itself as facing outward, across national, cultural and regional boundaries. Kenyanness is tied to colonial history, but the traumatic legacies of forced labor migration are reframed not as a trauma to work through, but as a resource for contemporary identities and politics. These texts imagine a national identity that is transnational, transregional and to an extent global. In the process, such texts render it increasingly difficult to frame writers and texts, but also memory discourses, in simplistic national or regional frameworks. These texts, memories, and identities seem unbound.

Transregional Mnemoscapes

What does this analysis mean for the present volume's aim to conceptualize and describe regions of memory? Which kinds of regional framework or grid can be applied to such unbound narratives?

As we have seen, remembering forced labor migration as represented by these two novels creates spatial connections—memory regions—across large parts of the world. Geographically distant locations are connected through mnemonic, imaginative and narrative activity. Relational remembering as evidenced by *Dust* and *Vikram Lall* renders such pre-existing connections visible and create transnational and transregional imaginaries. In the process, traumatic memories of slave trade and indentured labor are transformed into origin stories. Particularly in Vassanji's *Vikram Lall*, the legacies of colonialism become heritage, something to remember fondly and affectionately, and thus adds a teleological connection from colonialism to Afrasian communities in Kenya. Thus conceived, both texts allocate a specific position to the legacies of colonialism within the Kenyan national narrative: colonial oppression is not Kenya's "single story", to borrow the words from Nigerian writer Chimamanda Ngozi Adichie (2009, n.p.), but is integrated into the vast reservoir of stories that make up national identities. These texts are not invested in stripping colonialism of its violence and unlawfulness, or deny the pain it caused and causes, but the colonial legacies filtered through processes of remembering gain a different significance as resources for ambiguous national identities.

Remembering forced labor migration thus opens up a transregional imaginative space to where different communities' "histories of victimization come to intersect" while simultaneously allowing for these memories to be reframed as resources "to build new worlds out of the materials of older ones" (c.f. Rothberg 2009, 8). Moreover, leading cultural and literary scholars perceive colonialism not only as a precursor for modernity, or the watershed moment that ushered in modernity; they also identify forced mass migration as the signature experience of many *modernities* across the globe (c.f. Appadurai 1996; Gilroy 1993; Ngũgĩ wa Thiong'o 2014). Kenyan author and critic Ngũgĩ wa Thiong'o (2014, 54) argues that "globality as much as coloniality are constant features in the postcolonial even when the latter refers exclusively to those societies and peoples impacted by imperial colonialism". He calls our attention to the "worldliness" of such mobile communities and entangled literatures, referencing the sheer vastness of the spatial connections evoked by such entangled imaginations (2014, 56). The colony, thus conceived, becomes the repository for cosmopolitan identities (c.f. Appadurai 1996), but is far from the only identity resource available to such writers. As Mukoma wa Ngugi puts it

> To think about South-to-South relations is to enter a place of great intellectual vulnerability. Once we leave the relationship of, let's say, Africa and Europe via colonialism, the world suddenly becomes very vast, complicated, and scary as the knowledge of how just little we know settles in. Yet, this place that is just outside our comfort zone is a beautiful place to be in—it's a place of discovery of new ideas and seeing old ideas anew. (Mukoma wa Ngũgĩ 2012)

In this vein, Indian Ocean scholar Isabel Hofmeyr (2012) calls for a paradigmatic change in how we approach and analyze the legacies of colonialism by rendering an oceanic approach method. Hofmeyr perceives the Indian Ocean as a "space of intersection" that puts frameworks such as nation, area and region to the test (2012, 584). Such trajectories effectively reach beyond the "template of the nation state and area studies", thus calling for other critical frameworks than the national or the regional (2012, 584).

By studying the mnemonic trajectories across the ocean, rather than containing them in national contexts, a wide range of "lateral networks" becomes visible (c.f. Hofmeyr 2012, 584). Gaurav Desai calls our attention to the "racial hybridity of the Indian Ocean world" and the "representations of creolized cultures [that] resis[t] the convenient but ultimately reductive binaries that often accompany our accounts of African identities" (Desai 2010, 718). The Indian Ocean is a reservoir of various histories and identities, not only with regard to littoral communities and their relation to this ocean, but also concerning networks and agents who use this space as a realm to remember and imagine movements across this Afrasian Sea.

These claims resonate within Paul Gilroy's influential study *The Black Atlantic*, in which Gilroy identifies the Atlantic as this "intercultural and transnational formation" that demolishes "different nationalist paradigms for thinking about cultural history" (1993, 1). Gilroy describes the

Atlantic Ocean as a space in which "the inescapable hybridity and intermixture of ideas" are played out (1993, xi). Following the canonical routes of the transatlantic slave trade, Gilroy observes that those who left Africa and came to the West are deadlocked "between (at least) two great cultural assemblages, both of which have mutated through the course of the modern world that formed them and assumed new configurations" (1993, 1). Black people in the (anglophone) West thus remain "locked symbiotically in an antagonistic relationship marked out by the symbolism of colours" (ibid.). This gridlock is brought about by the rhetoric and practices of cultural nationalisms and "cultural absolutisms", and the logic of this symbolism of cultures establishes a direct connection to colonialism (1993, 3). Such a binary perspective blocks from view other histories, connections and identities that are not colonial in origin. What is more, it prioritizes looking at what happened to African emigration and subsequent immigration to the West, whilst ignoring the fundamental transformations that occurred in Africa itself (c.f. De Souza and Murdoch 2005). What is important for this analysis is the way in which Gilroy theorizes the Atlantic as a connective, hybrid and unbound space: "The history of the black Atlantic yields a course of lessons as to the instability and mutability of identities which are always unfinished, always in the making" (1993, xi). In this sense, Gilroy was the first scholar to call for and work out an oceanic perspective on cultural histories. By studying these entangled histories and identities across the oceans, other connections than the canonical South-North, colonizer-colonized paradigms become visible.

Hofmeyr's call for an oceanic perspective as method hence ties in with Gilroy's claims:

> Analytically the Indian Ocean obliges us to extend our axes of investigation and hence has implications for literary scholars. Not only does it present an ideal vantage point for pursuing productive postnational and post-area studies, it also offers rich ways to conceptualize lateral transnationalisms within a previously Third World space. It is hence an ideal arena from which to discuss transnationalisms within the Global South. (Hofmeyr 2012, 589)

Here, the Indian Ocean becomes the space in which to imagine and ultimately understand transnational imaginations, yet at the same time this stance upholds an established distinction: to perceive the Indian Ocean and the Atlantic Ocean as separate mnemonic spaces with their own historical legacies. There appears to be a separatist logic to these Indian and

Atlantic Ocean perspectives as well. To structure two separate mnemonic spaces or memory regions along the confines of these waters does not correspond to what the two novels discussed here represent. The case studies pertinent to this essay, *Vikram Lall* and *Dust*, if read in juxtaposition, show how the imagination of a Kenyan national identity connects histories and identities move across the Atlantic and the Indian Oceans; with regard to precolonial trade routes, old colonial forced migration patterns, and contemporary travel trajectories. What these texts show us is that processes of remembering and the situational actualization of these memories do not stop at the border between the Indian and Atlantic Oceans.

Towards a Transoceanic Perspective

The two novels analyzed here stand for literatures which in their attempts to imagine national identities reach beyond themselves. "Geopoetics", to borrow a phrase from Stephanos Stephanidis, are intertwined with "geopolitics" but in unconventional manners, and read histories and memories against the grain (2009, 197). Through acts of remembering, transoceanic connections are established and become visible. Such relational memory processes and their representation in works of fiction put established comparative frameworks to the test, as we have seen. The framework of Kenyan literature as national literature reaches its conceptual limits here, because the search for a national identity effectively entangles Kenyanness with India, Brazil, Uganda, Tanzania, England, and Germany. The two novels discussed here are best described not as works of Kenyan national literature, but as literature with an interest in exploring Kenyanness. Although this may seem like an obvious point, given the transcultural backgrounds of both authors discussed here, both texts arguably form part of a body of fictional works which "mark stylistic and thematic departures that deliberately undermine the nationalist [...] impulses that governed earlier [African] writing" (Mwangi 2009, 1). These texts pursue a twofold "geopoetic" trajectory—one facing inward in an attempt to, in the cases discussed here, explore the many facets of Kenyan history and national identity (Stephanidis 2009, 197). The other trajectory points outwards, underscoring "the desire of African societies to reach out to the rest of the world" (Mwangi 2009, 5). This literature *on* Kenya inscribes itself into broader discourses, histories and identities that reach across national, cultural and regional boundaries. The memories of forced labor migration are set in productive relation to histories that are marginalized by the dominant (post)colonial narrative, e.g. in the form of ancient trade routes cutting across and moving alongside colonial pathways. In an effort to "self-fashion" contemporary Kenyan national identity, both texts are invested in "expres[sing] and refigure[ing] different social facts" (Mwangi 2009, 9). In the process, these texts render visible South-South relations and challenge the implicit hierarchies, representational and identity-political, of Global North and Global South.

The multiple and often unexpected entanglements that these novels make visible can only be adequately theorized through a transoceanic lens. This fact has implications for both the comparative study of literature and the field of memory studies. A transoceanic paradigm for the comparative study of literature would allow for a productive juxtapositional reading of texts which are oftentimes confined to the aforementioned grids of Indian or Atlantic Ocean, not to mention national literature frameworks. Such a transoceanic perspective would allow for a critical analysis of the "worldliness" of these texts, thus contributing to the discussion on how to conceptualize literatures with a truly global reach. Vassanji's *Vikram Lall* and Owuor's *Dust* are not the only examples where such a transoceanic lens could enable different critical insights. Although a more thorough analysis is necessary, there are hints that transoceanic entanglements can be found in the texts of other authors. As early as 1901, Joseph Conrad's *Heart of Darkness* subtly problematizes the colonizer-colonized paradigm and creates a transoceanic connection between Africa as the "dark continent" and imperial London as the true "heart of darkness". Caryl Phillips' fictions create mnemonic networks connecting the Caribbean, Africa, America and England. His characters oftentimes very literally stretch across the oceans as they set out to make sense of their hybrid identities. Yvonne Adhiambo Owuor's most recent novel *The Dragonfly Sea* (2019) explores the historical bonds and ambiguous legacies created by Chinese exploration of East Africa in the fifteenth century.

As one can see here, such transoceanic texts have implications for the comparative study of memory, for a transoceanic perspective would allow for a juxtapositional reading of memory discourses across national and regional grids. In 2011, Astrid Erll identified the third phase of memory studies that is interested in conceptualizing memory processes as they cut across national and cultural borders. To study transoceanic memory discourses certainly would fall under that category. A transoceanic perspective would allow for a conceptual shift: even in their trans-dimension, nation and culture remain the core concepts while studying the movements and entanglements of memory. Both concepts in their singular form at some point became the driving forces of imperialism and colonialism, to maintain a sense of purity and authenticity of culture even when featured in anti-colonial struggles and decolonizing criticism (c.f. Gilroy 1993). A transoceanic approach to memory studies would enable us to identify other forms of collective organization, shared identities, and histories, without overly emphasizing nation and culture as analytical categories.

References

Achenbach, Ruth, Jan Beek, John Njenga Karugia, Rirhandu Mageza-Barthel, and Frank Schulze-Engler. 2020. Afrasian Transformations: An Introduction. In *Afrasian Transformations: Transregional Perspectives on Development Cooperation, Social Mobility and Cultural Change*, ed. Ruth Achenbach, Jan Beek, John Njenga Karugia, Rirhandu Mageza-Barthel, Frank Schulze-Engler, 1–17. Leiden: Brill.

Adichie, Chimamanda Ngozi. 2009. The Danger of a Single Story – TED Talk. https://www.ted.com/talks/chimamanda_ngozi_adichie_the_danger_of_a_single_story/transcript.

Allen, Richard Blair. 2014. *European slave trading in the Indian Ocean, 1500–1850.* Indian Ocean studies. Athens, Ohio: Ohio University Press.

Anyanzwa, James. 2019. East Africa's joint mega railway project at the crossroads. January 28. https://www.theeastafrican.co.ke/business/East-Africa-joint-mega-railway-project-at-the-crossroads/2560-4954932-6la96xz/index.html.

Appadurai, Arjun. 1996. *Modernity at Large: Cultural Dimensions of Globalization.* Minneapolis/London: University of Minnesota Press.

Araujo, Ana Lucia, ed. 2012. *Politics of Memory: Making Slavery Visible in the Public Space.* New York: Routledge.

Araujo, Ana Lucia, Mariana P. Candido, and Paul E. Lovejoy, eds. 2011. *Crossing Memories: Slavery and African Diaspora.* Trenton, NJ: Africa World Press.

Clark Hine, Darlene, ed. 2009. *Black Europe and the African Diaspora.* Urbana: University of Illinois Press.

Desai, Gaurav. 2010. Oceans Connect: The Indian Ocean and African Identities. *PMLA*, 125, No. 3: 713–20.

De Souza, Pascale, and H. Adlai Murdoch, 2005. Editorial Introduction – Oceanic Dialogues: From the Black Atlantic to the Indo-Pacific. *International Journal of Francophone Studies* 8, no. 2: 133–46.

Dubey, Ajay. 2010. *Indian Diaspora in Africa: A Comparative Perspective.* New Delhi: African Studies Association of India.

Eckstein, Lars. 2006. *Re-membering the Black Atlantic: On the Poetics and Politics of Literary Memory.* Amsterdam: Rodopi.

Edugyan, Esi. 2018. Washington Black. London: Serpent's Tail.

Erll, Astrid. 2011. Travelling Memory. *Parallax* 17, no. 4: 4–18.

———. 2018. Homer: A relational mnemohistory. *Memory Studies* 11, no. 3: 274–86.

Gilroy, Paul. 1993. *The Black Atlantic: Modernity and Double Consciousness.* Cambridge, MA: Harvard University Press.

Hofmeyr, Isabel. 2012. The Complicating Sea: The Indian Ocean as Method. *Comparative Studies of South Asia, Africa and the Middle East* 32, no. 3: 584–90.

Jayasuriya, Shihan de Silva, ed. 2003. *The African Diaspora in the Indian Ocean.* Lawrenceville: Africa World Press.

———. 2010. The African Diaspora in Asian Trade Routes and Cultural Memories. Lewiston: Mellen.

Killingray, David. 2001. African Voices from the Two World Wars. *Historical Research* 74, no. 186 (November): 426–43.

Klein, Herbert S., and Francisco Vidal Luna. 2010. *Slavery in Brazil.* Cambridge/New York: Cambridge University Press.

Morrison, Toni. 2007 [1987] . *Beloved.* New York: Random House.

Mwangi, Evan Maina. 2009. *Africa Writes Back to Self: Metafiction, Gender, Sexuality.* Albany: State University of New York Press.

Mwere, David. 2017. Kenyans of Indian descent become 44th tribe. July 22. https://www.nation.co.ke/news/Hindu-officially-becomes-Kenya-s-44th-tribe/1056-4027242-12m5ygj/index.html.

Ngugi, Mukoma wa. 2012. Rethinking the Global South. *The Global South Project*, n.p.

Njenga Karugia, John. 2018. Connective Afrasian Sea Memories: Transregional Imaginaries, Memory Politics, and Complexities of National "Belonging". *Memory Studies* 11, no. 3: 328–41.

Olaniyan, Tejumola, and James H. Sweet, eds. 2010. *The African Diaspora and the Disciplines.* Bloomington: Indiana University Press.

Owuor, Yvonne Adhiambo. 2014. *Dust.* New York: Alfred A. Knopf.

———. 2019. *The Dragonfly Sea.* New York: Alfred A. Knopf.

Rothberg, Michael. 2009. *Multidirectional Memory: Remembering the Holocaust in the Age of Decolonization.* Stanford: Stanford University Press.

Schulze-Engler, Frank. 2012. The Commonwealth Legacy. Towards a Decentered Reading of World Literature. In *Literature for Our Times: Postcolonial Studies in the Twenty-First Century*, ed. Bill Ashcroft, Ranjini Mendis, Julie McGonegal, and Arun Mukherjee, 3–14. Amsterdam: Rodopi.

Sen, Amartya. 2006. *Identity and Violence the Illusion of Destiny.* London: Penguin Books.

Stephanidis, Stephanos. 2009. Turning East. In *Literature for Europe?* ed. Theo D'haen, and Iannis Goerlandt, 197–211. Leiden: Brill

Thiong'o, Ngũgĩ wa. 2014. *Globalectics: Theory and the Politics of Knowing*. New York: Columbia University Press.
Vassanji, M. G. 2003. *The In-Between World of Vikram Lall*. New York: A.A. Knopf.

Afterword: The Discourse of Regions

Jeffrey K. Olick

The lexicon of the now well-established field of memory studies has grown vast enough to demand … if not a moratorium on new terms, then at least a strong case for any expansion. And this is perhaps especially the case with the proposed "regions of memory," since "region" contains ambiguities, both etymological and in the ways we have used it here.

In the first place, region can refer to both a portion of another, larger entity—such as the southwestern United States—as well as to an area that encompasses multiple entities—such as "the Baltics" or "Northeast Asia." Indeed, the latter sense is the one for which we used to more commonly employ the term "area," as in "area studies," though the fall of the latter from grace, or at least from institutional investment, has been underway for decades, given its affiliations with modernization theory and wider processes of Western othering. Although the chapters in this book do mostly refer to larger areas (if areas is in fact what they are), many of the processes they chart are also at work within nation-states and even, in the case of borderlands, between portions of two or more nation-states, in ways that remain peripheral to homogenizing national programs. In either

J. K. Olick (✉)
Department of Sociology and History, University of Virginia
Charlottesville, VA, USA
e-mail: jko3k@virginia.edu

S. Lewis et al. (eds.), *Regions of Memory*, Palgrave Macmillan Memory Studies, https://doi.org/10.1007/978-3-030-93705-8_10

case, however, the reference is colloquially to a geographical space, either a part of a larger space or a group of contiguous spaces. As we have seen in the forgoing essays, however, this connection of space to place—to land and area—is quickly and productively surpassed when the geographical becomes cultural or conceptual.

In the second place, the etymological history of the word region entails a reference to rule, similar to regime, from the Latin *regere*. A region, in this way, refers to a territory that is governed or administered. And yet, as Joanna Wawrzyniak and Simon Lewis write in their introduction to this volume, our first conceptualization of regions of memory, while referring to matters of governance, was meant to identify relations and processes that take place *between* or *among* states, rather than to control of populations and cultures within them. By the same token, thinking in terms or regions—even if they are comprised of groups of states—simultaneously reinforces and calls into question the personification of states as robust individuals governing their internal processes and divisions. The concept of regions of memory, then, ends up complexifying the original referent—ruling or governing—because it calls our attention to the fact that such governance is never entirely internal, that identities are never formed in isolation. The forms of governing, moreover—be they political, cultural, or otherwise—can have very long and not necessarily uninterrupted reaches.

To be sure, the identification of "regions of memory" did begin with a (perhaps falsely) straightforward observation. Namely, from the perspective of the emerging field of memory studies over the past twenty years, it seemed to us that discussion of the politics of memory, while often formally comparable in terms of processes and structures in many different places, was formed around different issues in different parts of the world, and that these could be understood in terms of geographical linkages between the level of individual states and the global or cosmopolitan. For instance, however global or universal a memory "the Holocaust" might entail as a referent, it is not quite as defining a topic of concern in, say, South Korea and Japan as it is in Central and East Central Europe; in the same way, while mnemonic topics like "comfort women" or "the Nanjing Massacre" might be well-known in European discourses, their salience in Northeast Asia is rather different. And so, one might begin a list of "regions of memory"—like Northeast Asia (in which the legacies of Japanese imperialism are a defining feature); Central and East Central Europe, or alternatively, what Timothy Snyder called "The Bloodlands" (in which the Holocaust and other forms of massacre are central referents); or in the "Southern Cone" of Latin America (in which the legacies of authoritarian regimes and topics like "Los Desaparacidos" are characteristic). There are distinct discourses about different issues in different parts of the world. One interesting implication of this recognition, however, is that regions are in many ways defined not just by geographical proximity, but by their involvement in these discourses about issues that bring them together topically just as they divide them perspectivally, whether or not they are geographically contiguous or even nearby.

Very quickly, however, it became clear—and the chapters in this volume push this realization forward in immensely creative and important ways—that the linkage between the spaces identified in this way with geography or area breaks down—or perhaps is broadened out—into conceptual and cultural linkages more general than contiguous geography or interstate diplomacy. In the first place, there are inter- and transnational areas that share similar pasts, whether or not they respond to them in similar ways (e.g. the similar experiences of, say, the Baltic or Caucasian states during the Soviet era, or of Central European states under German occupation, or, even more distinctly, Latin American states, each of which had its own dictatorship). In other words, there are different states that form regions because they had similar experiences, often because of their location; in this way, we can refer to clusters of states that have dealt with similar pasts that are often, but not always, near each other geographically, as having regional similarities. Second, though, these examples are slightly different from others in which the commonality is defined not by similar experiences, but by experiences with or implications in the *same* history. For instance, an historical problem can define a point of contention between or among states (once again, "comfort women" or Japanese imperialism in relations among South Korea, China, and Japan). Additionally, there are yet other regions—and here we move most clearly away from direct geographical reference—defined by a general problematic, like the settler colonialism that forms a conceptual region among those nations founded on it, or even by their involvement in geo-political periods like "the Cold War." In turn, such de-spatialized examples (for instance, settler colonialism that might define Australia, the U.S., and Canada as in the same conceptual region despite their geographical separation) lead to shared work, or different strategies, for dealing with specific issues, like indigenous dispossession or child removal. Finally, there are complex flows through conduits like language, literature, and concepts, like Francophone or Anglophone "regions," or "the postcolony/colonial" as cultural regions.

As a concept, then, "regions of memory" not only quickly burst the bounds of our original intent to locate memory discourses that take place between the level of the nation-state and the global and to tie these discourses to geographical locations (though, to be sure, many regions of memory are indeed tied to specific locations). So, for instance, geographical location need not be—though it can be—a matter of contiguity. Krishan Kumar's analysis of empires as regions of memory can be applied equally well to empires defined by the conquering or annexation of neighboring territories (land-based empires) as it can be to those defined by the acquisition of far-flung colonies (overseas empires); this geographical complexity is even clearer in Laura Pozzi's account of how regions are created by the conceptual and other work of powerful agents. In a different way, David Lowe and Tony Joel's chapter is a particularly powerful demonstration of how the meaning of a single historical referent—the Cold War—is both remembered differently in different regions of its effect, as well as that the cosmopolitan meaning of the referent is generated out of these different perspectives; something similar could be said of Katharine McGregor's analysis of the ways in which the memory of an event that takes place in one place is shaped by entities and processes in other places, and ultimately by concepts—e.g. human rights—whose meanings flow back and forth across spaces and cases. Dirk Moses's chapter untangles the complex meanings and operation of the concept of genocide within, between, and across local and transnational contexts. Finally, Hanna Teichler, Ann Rigney, and Ksenia Robbe demonstrate in powerful ways what happens when we view matters from the point of view of cultural rather than strictly administrative or political perspectives.

As Joanna Wawrzyniak and Simon Lewis acknowledge in their introduction, we could of course have included more chapters on different regions (and different understandings of region) than we did, though our hope is that the chapters presented here, and our introduction and afterword, will spur further use of and thinking about the regional and regions in the analysis of memory culture (and vice versa). Taken together, the foregoing chapters have shown in remarkably poignant ways that regions both shape our ways of remembering and are shaped by them. Northeast Asia or East Central Europe may not be constituted solely by the memories and institutions of memory of the relevant pasts mentioned above, but our understanding of them as regions—and of regions as salient constellations—would involve different associations were these issues not at stake within them. The concept of regions of memory thus reconfigures our understanding of the dynamics of memory, just as attention to the dynamics of memory reconfigures not only our understanding of regions, but the identification and operation of regions themselves. We thus look forward to the generative impact of these important essays in our understanding of memory within, across, and beyond nation states, and in our understanding of cultures and discourses as being in constant flux, though a flux organized by the effort to understand complex pasts.

Index[1]

[1] Note: Page numbers followed by 'n' refer to notes.

S. Lewis et al. (eds.), *Regions of Memory*, Palgrave Macmillan Memory Studies, https://doi.org/10.1007/978-3-030-93705-8

www.ingramcontent.com/pod-product-compliance
Lightning Source LLC
Chambersburg PA
CBHW071815061025
33644CB00006B/680
* 9 7 8 3 0 3 0 9 3 7 0 4 1 *